U.S. Marine Corps Drill and Ceremonies Manual

Department of Defense

Pentagon
Publishing

DEPARTMENT OF THE NAVY
HEADQUARTERS UNITED STATES MARINE CORPS
2 NAVY ANNEX
WASHINGTON, DC 20380-1775

MCO P5060.20
C 06
5 May 03

MARINE CORPS ORDER P5060.20

From: Commandant of the Marine Corps
To: Distribution List

Subj: MARINE CORPS DRILL AND CEREMONIES MANUAL

Ref: (a) Marine Corps Manual

Encl: (1) Locator Sheet

1. Purpose. The Marine Corps Drill and Ceremonies Manual is designed to provide uniformity and standardization for all Marine Corps organizations. This Manual prescribes procedures for all close order drill and military ceremonial evolutions.

2. Cancellation. NAVMC 2691.

3. Summary of Revision. This Manual has been revised to incorporate minor changes to existing procedures and rearrange material more logically. Additionally, a large amount of new information has been added to the Manual. This Manual should be reviewed in its entirety.

4. Scope. This Manual encompasses all close order drill procedures for use by Marine Corps organizations with the exception of Marine Barracks, Washington, DC. The ceremonies outlined in this Manual are those traditionally executed by Marine Corps units. The only deviation from procedures outlined in this Manual are those authorized by specific provisions of the Manual.

5. Recommendation of Modification. Recommendations for improving or modifying the Marine Corps Drill and Ceremonies Manual are invited. Recommended changes will be submitted to the CG, Training and Education Command (TECOM), Quantico, VA.

6. Action. COs and all personnel shall familiarize themselves with this Manual.

7. Reserve Applicability. This Manual is applicable to the Marine Corps Reserve.

8. Certification. Reviewed and approved this date.

EDWARD HANLON, JR.
By direction

DISTRIBUTION: PCN 10207170600

Copy to: 8145001 (1)

LOCATOR SHEET

Subj: MARINE CORPS DRILL AND CEREMONIES MANUAL

Location: ____
(Indicate the location(s) of the copy(ies) of this Manual.)

MARINE CORPS DRILL AND CEREMONIES MANUAL

CONTENTS

CHAPTER

MARINE CORPS DRILL AND CEREMONIES MANUAL

CONTENTS

MARINE CORPS DRILL AND CEREMONIES MANUAL

INTRODUCTION

One of the cornerstones of Marine Corps customs, courtesies and traditions is our execution of close order drill and ceremonies. These traditions are perpetuated from one generation to the next through constant use and practice. The "esprit de corps" of every Marine has been brought about by their ever-present feeling of pride, not only in their unit, but also in themselves. The preservation of traditional discipline, our customs and courtesies, and the heritage of our Corps is our duty. It is our further duty to see that the same high standard of discipline and esprit de corps is not only preserved, but also further strengthened. These intangibles must be passed on to the future Marines who will take their place among the ranks our Corps.

The object of close order drill is to teach Marines by exercise to obey orders and to do so immediately in the correct way. Close order drill is one foundation of discipline and esprit de corps. Additionally, it is still one of the finest methods for developing confidence and troop leading abilities in our subordinate leaders.

Ceremonial duties are written deep into our history as a Corps. It is no coincidence that among the units famous for ceremonial prowess and spit-and-polish are also to be found some of the worlds most redoubtable fighting formations. The hallmark of the world's fighting organizations, the Roman legions, the Spartans, the Foreign Legion, the British Brigade of Guards and many others is that they are as good on parade as they are in the field or in the attack. While we no longer use drill and formations to align the ranks as was done for the phalanxes of Rome or the squares of Waterloo, drill and ceremony is still the foundation of instilling and developing discipline in any size unit and the individual.

The Marine Corps has long enjoyed a worldwide reputation for sharp appearance and soldierly performance of duties. Our reputation as a Corps is built upon the legacy of those that came before us and puts us on par with the finest military organizations in the world's history. This reputation has been enhanced by continually demonstrating that our execution of peacetime functions is excelled only by our performance in battle.

The special distinction that the Marine Corps has today in regards to drill and ceremonies is best exemplified by the saying *"First on foot, and right on the line."* This privilege bestowed on the Corps by the Secretary of the Navy on 9 August 1876, gives Marines the place of honor in any Naval formation.

Today we continue the tradition of excellence associated with drill. Beginning with OCS and recruit training we continue to use close order drill as the foundation for developing discipline and esprit de corps. This process is the first step that transforms the civilian into a Marine and then develops the teamwork and unit cohesion. This process then continues in the operating forces through the proper execution of drill and ceremonies to maintain the same level of discipline and esprit de corps found at entry-level training.

This Manual has been the work of many dedicated Marines, past and present. The members of the current drill committee owe all those who have contributed to this collaborative effort a debt of gratitude. Because of this the members of the drill committee in reviewing and updating this Manual have attempted to remain true to this heritage and the traditions of our Corps. This is the first drill manual of the 21st century and we have attempted to provide for the needs of a modern Marine Corps while remaining faithful to the traditions of our past. Extensive research was done to ensure historical accuracy for terms and procedures. Changes were made only to standardize usage and in some cases ensure compliance with other Marine Corps directives. Finally, the intent was to produce a manual that would be easy to use.

MARINE CORPS DRILL AND CEREMONIES MANUAL

PART I: DRILL

CHAPTER 1

INTRODUCTION TO DRILL

FIGURE

TABLE

MARINE CORPS DRILL AND CEREMONIES MANUAL

PART I: DRILL

CHAPTER 1

INTRODUCTION TO DRILL

1000. GENERAL

1. The Marine Corps Drill and Ceremonies Manual is designed to provide uniformity and standardization for all Marine Corps organizations. This Manual prescribes procedures for all close order drill and military ceremonial evolutions. Commanders will only deviate from prescribed procedures when expressly authorized for specific provisions of this Manual. The use of ceremonial drill movements authorized for Marine Barracks, Washington DC are not authorized for any other Marine Corps organization.

2. Online Access. In addition to the information found in this Manual, organizations and individuals may access the Marine Corps Drill web site by linking from Marine LINK at http://www.usmc.mil/. This web site not only contains an electronic version of this Manual but also sample LOIs for the various ceremonies described in this Manual.

3. Purpose of Drill. Commanders use drill to:

a. Move units from one place to another in a standard, orderly manner.

b. Provide simple formations from which combat formations may be readily assumed.

c. Teach discipline by instilling habits of precision and automatic response to orders.

d. Increase the confidence of junior officers and noncommissioned officers through the exercise of command, by the giving of proper commands, and by the control of drilling troops.

e. Give troops an opportunity to handle individual weapons.

4. Purpose of Formations

a. To build unit cohesion and esprit de corps by recognizing Marines during awards and promotion ceremonies.

b. To maintain continuous accountability and control of personnel.

c. To provide frequent opportunities to observe the appearance and readiness of the uniforms, arms, and equipment of the individual Marine.

d. To keep the individual Marine informed by providing the means to pass the word.

e. To develop command presence in unit leaders.

f. To instill and maintain high standards of military bearing and appearance in units and in the individual Marine.

g. To add color and dignity to the daily routine by reinforcing the traditions of excellence associated with close order drill.

5. Scope. This Manual prescribes drill for general use. Diagrams shown must be adapted to the strength of the unit and available space. Unless otherwise stated, whenever the masculine or feminine gender is used, both men and women are included.

1001. DEFINITIONS. The following definitions are standard terms used throughout this Manual.

1. Alignment. The dressing of several elements on a straight line.

2. Assembly Area. A designated location for forming units of platoon size or larger in preparation for a parade, review or ceremony.

3. Arms. A term used to normally designate the service rifle but can refer to any weapon. When in formation and a mix of weapons is carried the term arms will be used to designate all types of weapons.

4. Base. The element on which a movement is regulated.

5. Cadence. A rhythmic rate of march at a uniform step.

6. Center. The middle element of a formation with an odd number of elements (See figure 1-1a) or the left center element of a formation with an even number of elements. (See figure 1-1b.)

Figure 1-1a.--Center (Odd Number of Elements), Depth, and Front.

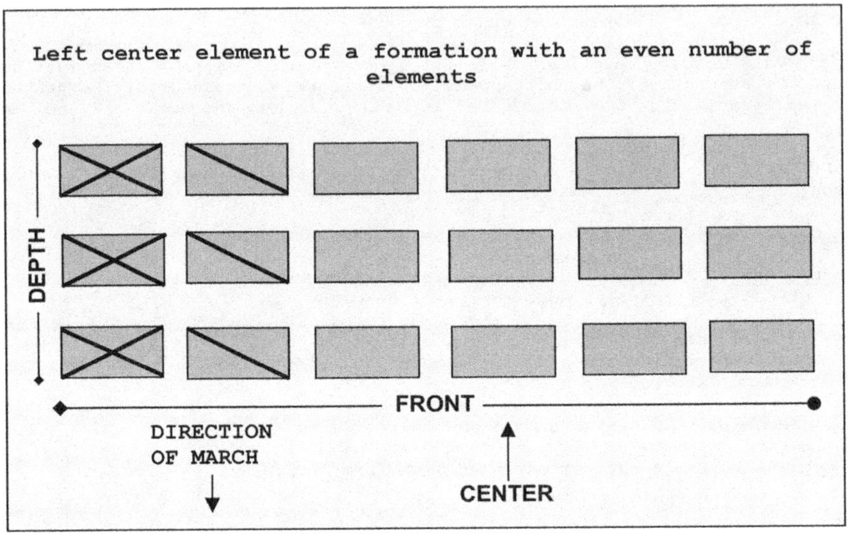

Figure 1-1b.--Center, Even Number of Elements.

7. Ceremony. A formal military formation designated to observe a specific occasion.

8. Column. A formation in which elements are placed one behind the other. A section or platoon is in column when members of each squad are one behind the other with the squads abreast of each other.

9. Commander of Troops (COT). The COT is the senior officer taking part in the ceremony. If an enlisted ceremony the COT is the senior enlisted.

10. Depth. The space from head to rear of an element or a formation. (See figure 1-1a.) The depth of an individual is considered to be 12 inches.

11. Distance. The space between elements in the direction of depth. Between individuals, the space between your chest and the person to your front. Between vehicles, the space between the front end of a vehicle and the rear of the vehicle to its front. Between troops in formation (either on foot, mounted, or in vehicles), the space from the front of the rear unit to the rear of the unit in front. Platoon commanders, guides, and others whose positions in a formation are 40 inches from a rank are, themselves, considered a rank. Otherwise, commanders and those with them are not considered in measuring distance between units. The color guard is not considered in measuring distance between subdivisions of the unit with which it is posted. In troop formations, the distance between ranks is 40 inches. (See figure 1-2.)

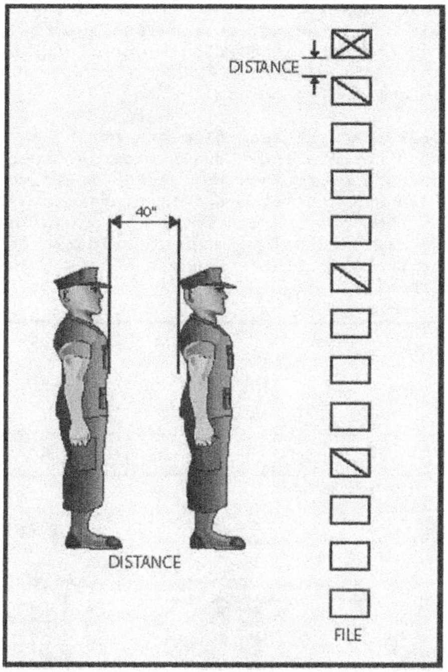

Figure 1-2.--Distance and File.

12. Double Time. Cadence at 180 steps (36 inches in length) per minute.

13. Element. An individual, squad, section, platoon, company, or other unit that is part of a larger unit.

14. Extended Mass Formation. The formation of a company or larger unit in which major elements are in column at close or normal interval and abreast at a specified interval greater than normal interval.

15. File. A single column of troops or vehicles one behind the other.

16. Flank. The right or left extremity of a unit, either in line or in column. The element on the extreme right or left of the line. A direction at a right angle to the direction an element or a formation is facing.

17. Formation. Arrangement of elements of a unit in line, in column, or in any other prescribed manner.

18. Front. The space occupied by an element or a formation, measured from one flank to the other. (See figure 1-1a.) The front of an individual is considered to be 22 inches.

19. Guide. The individual (base) upon whom a formation, or other elements thereof, regulates its march. To guide: to regulate interval, direction, or alignment; to regulate cadence on a base file (right, left, or center).

20. Head. The leading element of a column.

21. Interval. The lateral space between elements on the same line. (See figure 1-3.) Interval is measured between individuals from shoulder to shoulder and between vehicles from hub to hub or track to track. It is measured between elements other than individuals and between formations from flank to flank. Unit commanders and those with them are not considered in measuring interval between elements of the unit. Normal interval between individuals is one arm's length. Close interval is the horizontal distance between shoulder and elbow when the left hand is placed on the left hip.

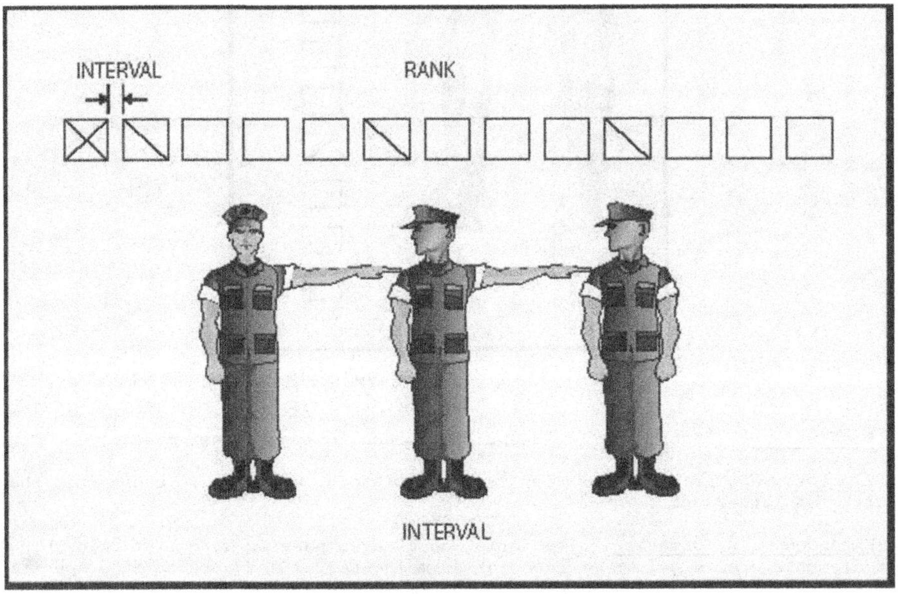

Figure 1-3.--Interval and Rank.

22. Left (Right). Extreme left (right) element or edge of a body of troops.

23. Line. A formation in which the elements are side by side or abreast of each other. A section or platoon is in line when its squads are in line and one behind the other.

24. Line of March. The line on which individuals or units are to march on.

25. Line of Troops. The line on which troops are to form when in formation.

26. Loosened Sling. Indicates a sling adjusted for the movement sling arms.

27. Mass Formation. The formation of a company or larger unit in which the major elements are in column at close interval and abreast at close interval.

28. Muffling. The procedure of draping colors for mourning with a mourning streamer or black bunting. It also refers to the process of muffling the musical instruments of a band for specific types of ceremonies.

29. Pace. The length of a full step in quick time, 30 inches.

30. Parade. A parade is a ceremony that involves the movement of marching units.

31. Parade Sling. A sling that has all excess slack removed and is taught. The keeper is adjusted and locked in a position next to the sling tip. The sling lies on the left side of the rifle.

32. Piece. An individual firearm such as a rifle.

33. Point of Rest. The point toward which all elements of a unit establish their dress or alignment.

34. Quick Time. Cadence at 112 to 120 steps (12, 15, or 30 inches in length) per minute. It is the normal cadence for drills and ceremonies.

35. Rank. A line of troops or vehicles placed side by side.

36. Review. A review is a type of ceremony that omits certain elements found in a parade, but includes an inspection (trooping the line) not found in a parade.

37. Rigged. This term refers to the condition when uniforms and equipment are properly fitted out in the manner for which they were intended for use. Swords are considered rigged when attached to the frog (noncommissioned officers) or sword sling (commissioned officer). A Marine is rigged when wearing the prescribed uniform or equipment.

38. Slow Time. Cadence at 60 steps per minute. Used for funerals only.

39. Snap. In commands or signals, the quality that inspires immediate response. In drill the immediate and smart execution of a movement.

40. Step. The distance from heel to heel between the feet of a marching individual. The half step and back step are 15 inches. The right and left steps are 12 inches. The steps in quick and double time are 30 and 36 inches, respectively.

41. Strong Grip. The strong grip is when the thumb is wrapped around the front of the staff with the fingers wrapped to the rear. (See figure 1-4.)

42. Unit Leader. Is the individual who is drilling the unit. This can be any individual who is conducting drill or can be those assigned a specific billet such as squad leader, platoon sergeant, platoon commander, etc.

43. "V" Grip. The "V" grip is with the staff placed in the "V" formed by the thumbs and forefinger with the fingers extended and joined. (See figure 1-5.)

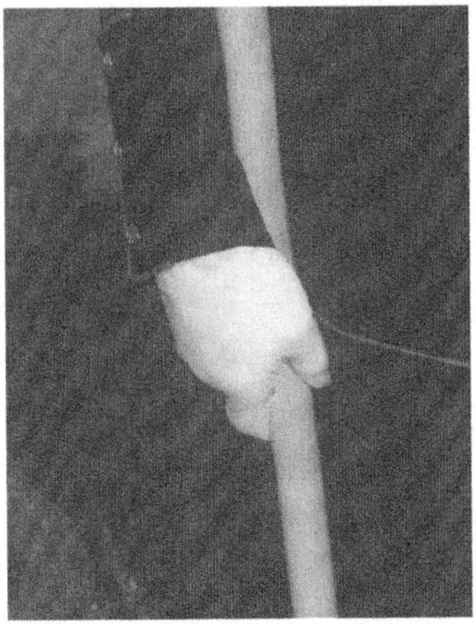

Figure 1-4.--Strong Grip on the Staff.

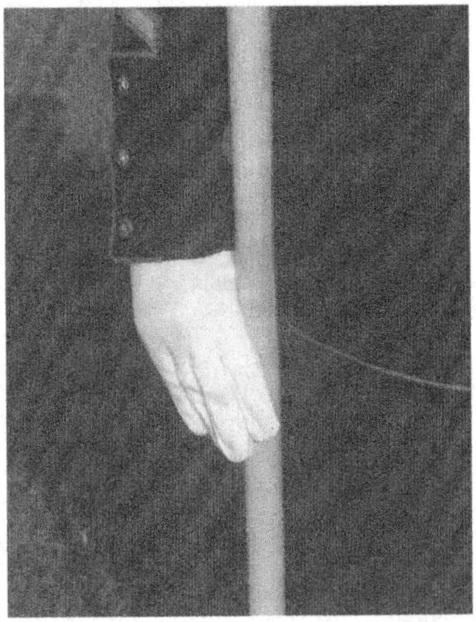

Figure 1-5.--"V" Grip on the Staff.

1002. INSTRUCTING DRILL

1. Training personnel in drill is an important duty. From the beginning of their career, officers and noncommissioned officers should take pride in being considered efficient drill instructors. If they know the drill regulations and how to instruct, they will gain the respect and confidence of those they command.

2. In addition, good drill instructors must:

a. Follow regulations strictly, as an example to the personnel under instruction.

b. Have energy, patience, and spirit.

c. Have military neatness and bearing.

d. Observe personnel closely, immediately correcting any mistakes noted.

3. When instructing drill movements the system described by the acronym P.C.W.C. will be used to ensure all required information is provided. P.C.W.C. stands for:

(P) Purpose of the movement.

(C) Number of Counts involved in its execution.

(W) When the movement may be executed.

(C) Commands to cause the movement to be executed.

4. When instructing drill movements for unit movement the system described by the acronym P.F.W.C will be used to ensure all required information is provided. P.F.W.C. stands for:

(P) Purpose of the movement.

(F) Formation from which the movement is executed.

(W) When the movement may be executed.

(C) Commands to cause the movement to be executed.

5. When the instructor corrects a movement, the individual or unit should immediately be required to repeat the movement properly.

6. Before a drill period, an instructor should thoroughly study the movements to be executed.

7. Instructors may place themselves wherever they can best control the troops, make corrections, and ensure proper performance.

8. The instructor briefly explains and demonstrates each new movement prior to its execution by the troops. The troops should take proper positions unassisted. Each position or movement must be thoroughly understood before another is attempted.

9. Drill periods should be short but frequent. Snap should be required in every movement.

1003. INSTRUCTION GROUPS

1. The basic instruction group is the squad. Its size facilitates individual instruction.

2. Individuals who learn slowly should be placed in special squads. The best instructors available should drill these squads.

1004. COMMANDS AND THE COMMAND VOICE

1. There are four types of commands: preparatory commands, commands of execution, combined commands, and supplementary commands. All commands in this Manual are shown in quotation marks, e.g., "**Present, ARMS**" and **bold print**.

a. The preparatory command indicates a movement is to be made and may also indicate the direction of the movement. In this Manual, preparatory commands are shown beginning with a capital letter followed by lower case letters. The comma indicates a pause between the preparatory command and the command of execution. Examples would be "**Forward,**" "**Left,**" "**Platoon,**" "**About,**" etc.

b. The command of execution causes the desired movement to be executed. In this Manual, commands of execution are shown in **CAPITAL LETTERS**. Examples would be "**MARCH,**" "**FACE,**" "**ATTENTION,**" etc.

c. With the combined command, the preparatory command and the command of execution are combined. In this Manual combined commands are shown in **UNDERLINED CAPITAL LETTERS**. Examples would be "**AT EASE**," "**REST**," "**FALL IN**," etc.

d. Supplementary commands are commands that cause the component units to act individually. An example would be the commands squad leaders would give to their individual squads following the platoon commander's preparatory command, "**Column of Files From the Right,**" and before the command of execution "**MARCH.**" In this Manual supplementary commands may be shown as preparatory commands, commands of execution or combined commands, depending on the movement.

2. When giving commands, commanders face their troops.

a. For company formations or larger, when commanding marching troops from the head of a column or massed formations, commanders march backward while giving commands.

b. When commanding a unit that is part of a larger unit, commanders turn their heads to give commands, but do not face about except when the unit is halted and the smaller units are in line. In this case, the leader faces about to give all commands except to repeat preparatory commands, for which turning the head is sufficient.

3. Commanders of platoons and larger units, when drilling as a part of a still larger unit, repeat all preparatory commands or give the proper new command or warning. There are three exceptions to this.

a. The first is that no repetition is necessary for combined commands such as "**FALL IN**," "**FALL OUT**," "**REST**," or "**AT EASE**."

b. The second is that no repetition of command is necessary when a unit is in mass formation.

c. The third exception is that no repetition of command is necessary during parades and ceremonies where the commander of troops, adjutant, etc., may be clearly heard by all hands or the commander of troops and adjutant, give combined commands and subordinate unit commanders cause their units to execute the command independently. (e.g., when the regimental commander gives the combined command to "**PORT ARMS**" subordinate battalion commanders would give the command of "**Port, ARMS**" so that their battalion would execute the command as an independent unit of the regiment.)

4. If at a halt, the commands for movements, which involve marching at quick time in a direction other than to the direct front, such as "**Column Right, MARCH,**" are not prefaced by the preparatory command, "**Forward.**"

5. The only commands that use unit designations, such as "Battalion" or "Company," as preparatory commands are "**ATTENTION**" and "**HALT.**" Such commands shall have no further designation added (e.g., "First Battalion, ATTENTION" or "Company C, HALT"). Commands shall be given only as stated herein.

6. A command must be given loud enough to be heard by all members of a unit.

a. Good posture, proper breathing, and the correct use of throat and mouth muscles help develop a commander's voice.

b. Projecting the voice enables one to be heard at maximum range without undue strain. To project a command, commanders must focus their voices on the most distant individuals. Good exercises for voice projection are:

(1) Yawning to get the feel of the open mouth and throat.

(2) Counting and saying the vowel sounds "oh" and "ah" in a full, firm voice.

(3) Giving commands at a uniform cadence, prolonging each syllable.

(4) When practicing, stand erect, breathe properly, keep the mouth open wide, and relax the throat.

c. The diaphragm is the most important muscle in breathing. It is the large horizontal muscle that separates the chest from the abdomen. It automatically controls normal breathing, but must be developed to give commands properly. Deep breathing exercises are one good method of developing the diaphragm. Another is to take a deep breath, hold it, open the mouth, relax the throat muscles, and snap out a series of fast "hats" or "huts." Expelling short puffs of air from the lungs should make these sounds. If properly done, you can feel the stomach muscles tighten as the sounds are made.

d. The throat, mouth, and nose act as amplifiers. They give fullness to and help project the voice. In giving commands, the throat should be relaxed. The lower jaw and lips should be loose. The mouth should be open wide and the vowel sounds (a, e, i, o, u) should be prolonged. Consonants (letters other than vowels) and word endings should be curt and sharply cut off.

e. The position of attention is the proper position for giving commands (See figure 1-6a). A leader's bearing will be emulated. If it is military, junior personnel will be inspired to respond to commands with snap and precision.

f. Distinct commands inspire troops. Indistinct commands confuse them. All commands can be given correctly without loss of effect or cadence. To give distinct commands, you must emphasize enunciation; make full use of the tongue, lips, and lower jaw; practice giving commands slowly, carefully, and in cadence; and then increase the rate of delivery until the proper rhythm (112 to 120 beats per minute) is reached and each syllable is distinct. Raising the hand to the mouth to aid in projecting commands is not proper.

g. Inflection is the rise and fall in pitch, the tone changes of the voice.

(1) Preparatory commands should be delivered with a rise and inflection in the voice. (e.g., "BaaaTAL ion," "PlaaaTOON," "FooreWARD," "TO the REAR," etc.) In particular those preparatory commands that cause supplemental movements should be heavily accentuated on the last syllable. (e.g., The command **"Present, ARMS"** the preparatory command Preee(pause) ZENT" causes those armed with swords to execute the first count of the movement and the national color to go to the carry. Another example is **"Officers, Center, MARCH."** On the preparatory command of "OffiCERS" those armed with swords go to the carry, on the preparatory command of "CennnTER" the officer's step and/or face)

(2) A command of execution is given in a sharper and higher pitch than the tone of the preparatory command's last syllable. A good command of execution has no inflection, but it must have snap. It should be delivered with sharp emphasis, ending like the crack of a whip. If properly given, troops will react to it with snap and precision.

(3) Combined commands such as "**FALL IN**" are delivered without inflection. They are given in the uniform high pitch and loudness of a command of execution.

a. Position of Attention. b. Right Shoulder Arms.

Figure 1-6.--Correct Position for Giving Commands.

1005. GENERAL RULES FOR DRILL

1. Certain drill movements may be made toward either flank. This Manual explains such movements in one direction only. To move to the other direction, it is necessary to substitute the word "left" for "right" as shown in parentheses.

2. The command, "**AS YOU WERE**," cancels a movement or order started but not completed. At this command, troops should resume their former positions.

3. While marching, alignment is maintained toward the right, except:

a. Upon the command "**GUIDE LEFT**" or "**GUIDE CENTER**," alignment is maintained toward the left or center until the command "**GUIDE RIGHT**" is given.

b. Regardless of the direction in which alignment is established, at the command of execution for a drill movement involving marching, the direction of alignment is toward direction of the movement. (e.g., if the command of "**Column Left, MARCH**" is given, the unit aligns to the left until the column movement is completed.) Upon completion of the drill movement, alignment will be in the direction established prior to commencing the movement.

c. In a column of platoons, the guide of the leading platoon controls direction and distance, and maintains the cadence counted by the commander. Guides of the following platoons control distances and follow the direction and cadence set by the leading guide.

4. Slight changes in direction are made by:

a. Adding "**Half**" to the preparatory command for turning or column movements; for example, "**Column Half Right (Left), MARCH**," changes direction 45 degrees.

b. At the command "**INCLINE TO THE RIGHT (LEFT)**," guides change their direction of march slightly to the right (left).

5. Platoons in a company and individuals in a squad are numbered from right to left in line, and from front to rear in column. Squads in a platoon are numbered from front to rear in line and from left to right in column.

6. Posts of officers, noncommissioned officers, guidon bearers, and special units are shown in figures or explained later in this Manual.

a. Changes of post are made by the most direct route unless specifically prescribed. The movement is made as soon as possible after the command of execution. Persons who have duties in connection with the movement take their new posts after those duties have been performed. Individuals moving from one post to another or aligning units must maintain military bearing. All necessary facings and changes of direction must be made as if drilling.

b. After the initial formation, guidon bearers and special units maintain their positions with respect to the flank or end of the organization with which they were originally posted.

c. All officers or noncommissioned officers commanding armed troops will be armed. When giving commands, drilling troops, or making reports, the rifle will be carried at right shoulder arms (See figure 1-6b), the sword will be carried at the carry sword position, and the pistol will be carried in the holster. Facing movements by commanders of armed troops may be executed while at right shoulder arms.

7. Whenever drill movements are executed while troops are marching, the command of execution is given as the left foot strikes the deck if the movement is to the left, and as the right foot strikes the deck if the movement is to the right. The one exception is when executing left shoulder arms from port arms; the command of execution will be given on the right foot. (See table 1-1 for the suggested foot to give commands on while marching.)

8. The cadence of commands should be that of quick time, even though no marching is involved. The pause between preparatory commands and commands of execution should be adapted to the size and formation of the unit.

a. The best pause to allow for a squad or platoon while marching is one count or step between the two commands. (See table 1-1 for the suggested foot to give commands on while marching.)

(1) To develop the proper cadence for commands at a halt, count ONE, TWO, THREE, FOUR in quick time. Then give commands, such as "**Right, FACE,**" without interrupting the cadence:

ONE	TWO	THREE	FOUR
"**Right,**	(Pause)	**FACE**"	

(2) Marching commands, such as "**By the Right Flank, MARCH,**" must be started so the preparatory command will end as the foot in the desired direction of movement strikes the deck. There is then a full count before the command of execution, which is given on the same foot. (See table 1-1 for the suggested foot to give commands on while marching.)

b. For a company or larger unit, the intervals must be longer. This is necessary for leaders of component units to repeat preparatory commands, give warning, or supplementary commands. The following example shows the proper cadence for the command "**Right, FACE**" to a company:

ONE	TWO	THREE	FOUR
(Company Commander) "**Right,**	(Platoon Commander) {"**Right**"}		(Company Commander) **FACE**"

1006. DRILL BY THE NUMBERS

1. Drill movements may be divided into individual motions for instruction. When drill is executed by the numbers, the first motion is made on the command of execution. Subsequent motions are made in proper order on the commands TWO, THREE, FOUR, the number of counts depending upon the number of motions in the movement. To use this method, the command "**BY THE NUMBERS**" precedes the preparatory command. All movements are then executed by the numbers until the command "**WITHOUT NUMBERS**" is given.

RIGHT	LEFT	RIGHT	LEFT	RIGHT	LEFT	WHEN GIVEN
		"Squad,	(Step)	HALT"		From column or flank
	"Pla	toon,	(Step)	HALT"		From column or flank
	"Com pa ney		(Step)	HALT"		From column or flank
	"Right	Turn,	(Step)	MARCH"		While marching in company mass the color guard or staff
"Left	Turn,	(Step)	MARCH"			" " " " " " " " "
"Counter	March,	(Step)	MARCH"			For color guards.
"Mark	Time, "Mark	(Step) Time,	MARCH" (Step)	 MARCH"		May be given as either foot strikes the deck from line, column or oblique
"Half	Step, "Half	(Step) Step,	MARCH" (Step)	 MARCH"		May be given as either foot strikes the deck from line, column or oblique
"For	ward,	(Step)	MARCH"			From mark time or half step in column or flank
"By the Right		Flank,	(Step)	MARCH"		From column or flank
	"By the Left		Flank,	(Step)	MARCH"	From column or flank
RIGHT	**LEFT**	**RIGHT**	**LEFT**	**RIGHT**	**LEFT**	**WHEN GIVEN**
	"Column	Right,	(Step)	MARCH"		From column
"Column	Left,	(Step)	MARCH"			From column
"Column	Half	Right,	(Step)	MARCH"		From column
	"Column	Half	Left,	(Step)	MARCH"	From column
	"To the Rear,		(Step)	MARCH"		From column or flank
	"Right O blique,		(Step)	MARCH"		From column
"For	ward,	(Step)	MARCH"			From right oblique
"Pla	toon,	(Step)	HALT"			From right oblique
	"In	Place,	(Step)	HALT"		To halt in the right oblique while marching at quick time, mark time or half step.
"Re	sume,	(Step)	MARCH"			To resume marching in the left/right oblique from in place halt, mark time or half step
"Left O blique,		(Step)	MARCH"			From column
	"For	ward,	(Step)	MARCH"		From left oblique
	"Pla	toon,	(Step)	HALT"		From left oblique
	"In	Place,	(Step)	HALT"		To halt in the left oblique while marching at quick time, mark time or half step.

Table 1-1.--Suggested Foot to Give Marching Commands On.

RIGHT	LEFT	RIGHT	LEFT	RIGHT	LEFT	WHEN GIVEN
"Close,	(Step)	MARCH"				To close the interval in column marching
"Ex	tend,	(Step)	MARCH"			To extend the interval in column marching
"Count	Cad	ence,	(Step)	COUNT"		From column
"De ed	lay	Cad	ence,	(Step)	COUNT"	From column
COMMANDS FOR THE MANUAL OF ARMS WHILE MARCHING						
"Port,	(Step)	ARMS"				From left shoulder
"Left	Shoul	der,	(Step)	ARMS"		From port arms
	"Left	Shoul	der,	(Step)	ARMS"	From right shoulder
"Right	Shoul	der,	(Step)	ARMS"		From left shoulder or port
	"Port,	(Step)	ARMS"			From right shoulder
RIGHT	**LEFT**	**RIGHT**	**LEFT**	**RIGHT**	**LEFT**	**WHEN GIVEN**

Table 1-1.--Suggested Foot to Give Marching Commands On--Continued.

1007. MASS COMMANDS AND INDIVIDUAL COMMANDS FROM RANKS

1. Mass Commands. The use of mass commands in drill develops confidence and team spirit. It also teaches troops to give and execute commands properly.

a. The initial command is **"At Your Command."** The instructor then gives a preparatory command that describes the movement, for example, **"At Your Command, Face the Platoon to the Right, COMMAND."** After this, all members of the platoon command, **"Right, FACE"** together and execute it. Another mass command example is **"At Your Command, Call the Platoon to Attention, COMMAND."** The troops command, **"Platoon, ATTENTION."**

b. Marching movements may be conducted in a similar manner as follows: **"ALL MOVEMENTS UNTIL FURTHER NOTICE WILL BE AT YOUR COMMAND."**

(1) INSTRUCTOR: **"Call the Platoon to Attention, COMMAND."** TROOPS: **"Platoon, ATTENTION."**

(2) INSTRUCTOR: **"Forward, COMMAND."** TROOPS: **"Forward, MARCH."**

(3) INSTRUCTOR: **"By the Right Flank, COMMAND."** TROOPS: **"By the Right Flank, MARCH."**

c. Only simple movements with short preparatory commands are used for mass commands. Preparatory commands that require supplemental commands by subordinate leaders are not used.

d. In order to return to normal drill methods, the instructor orders **"At My Command."**

2. Individual Commands From Ranks. As an aid in training individuals to give commands properly, personnel in ranks may be designated to give individual commands. This should be done only after a unit has learned to execute commands and give mass commands. The instructor designates the individual who is to give the command by name or place in ranks. He may be designated while the unit is

marching or halted. The instructor describes the movement to be made, as in mass commands, but does not add "**COMMAND.**" The designated individual remains in position and gives the appropriate command, executing the command with the unit.

3. Counting Cadence. Counting cadence by troops in ranks teaches them group coordination and rhythm. The command is "**Count Cadence, COUNT**" or "**Delayed Cadence, COUNT.**" The command of execution being given as the left foot hits the deck. Everyone then counts cadence for 8 steps, starting when the left foot next strikes the deck. They should be required to do so in a firm, vigorous manner.

"Count Cadence, COUNT"

Step	LEFT	RIGHT	LEFT	RIGHT	LEFT	RIGHT	LEFT	RIGHT
Count	One	Two	Three	Four	One	Two	Three	Four

"Delayed Cadence, COUNT"

Step	LEFT	RIGHT	LEFT	RIGHT	LEFT	RIGHT	LEFT	RIGHT	LEFT	RIGHT
Count	One		Two		Three		Four		One	
	Two		Three		Four		One	Two	Three	Four
	One	Two	Three	Four	(United)	(States)	(Marine)	(Corps)		

1008. ADVANCED DRILL PROCEDURES

1. Cadence Drill (Speed Drill). Cadence drill is an advanced form of drill. It is used only with Marines who have learned basic drill as prescribed in this Manual. It provides variety for well-trained troops and "livens up" a drill period. In cadence drill, the commander, abbreviates preparatory commands and deletes the pause between the preparatory command and the command of execution. Each time a commander uses this form of drill, he must explain that the unit is departing from prescribed drill. Cadence drill is suitable for platoon and smaller units. Examples of cadence drill are:

Count One	Two	Three	Four	One
Step	LEFT	RIGHT	LEFT	RIGHT
Command (s)		**"Right**	**Flank**	**MARCH"**
	"Left	**Flank**	**MARCH"**	
	"Rear	**MARCH"**		

2. Trick Drill. As with cadence drill, this form of drill is used only with troops who have thoroughly learned all prescribed drill and cadence drill. Additionally, it may be used as a motivating device for well-trained troops. It is extremely complex and requires extensive training and rehearsals. Trick drills are not described in this Manual. They are limited only by the imagination of the drill instructor.

1009. TABLE OF SYMBOLS USED IN THIS MANUAL

1. Table 1-2 shows the symbols used in this Manual.

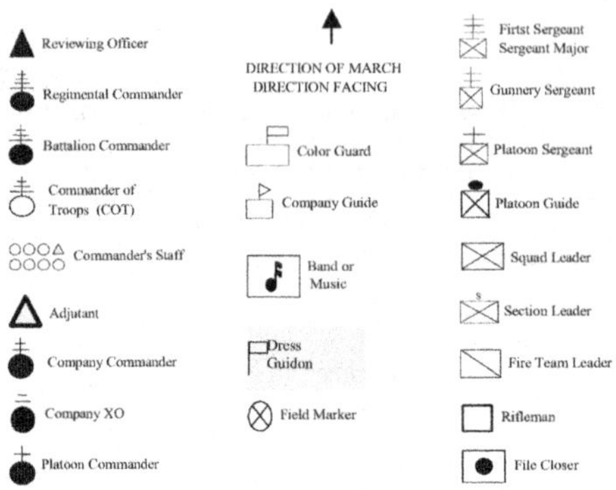

Table 1-2.--Drill Symbols.

MARINE CORPS DRILL AND CEREMONIES MANUAL

PART I: DRILL

CHAPTER 2

INDIVIDUAL INSTRUCTIONS WITHOUT ARMS

PAGE

FIGURE

MARINE CORPS DRILL AND CEREMONIES MANUAL

PART I: DRILL

CHAPTER 2

INDIVIDUAL INSTRUCTIONS WITHOUT ARMS

2000. GENERAL

1. Individual movements form the basis for building unit formations and movements.

2. Movements that can be executed from both the halt and while on the march should be explained, demonstrated, and practiced from the halt first.

SECTION 1: POSITIONS

2100. ATTENTION. The position of attention (see figure 2-1) is the basic military position from which most other drill movements are executed. There are no counts, however, there are seven steps in describing the position:

1. Smartly bring your left heel against the right.

2. Turn your feet out equally to form an angle of 45 degrees. Keep your heels on the same line and touching.

3. Your legs should be straight, but not stiff at the knees.

4. Keep your hips and shoulders level and your chest lifted.

5. Your arms should be straight, but not stiff at the elbows; thumbs along the trouser seams, palms facing inward toward your legs, and fingers joined in their natural curl.

6. Keep your head and body erect. Look straight ahead. Keep your mouth closed and your chin pulled in slightly.

7. Stand still and do not talk.

8. The movement may be executed when halted, at any position of rest, or while marching at route step or at ease. The commands are "**FALL IN**" and "**ATTENTION.**" On the command "**FALL IN,**" you would assume your position in ranks at the position of attention. The command "**ATTENTION**" is always preceded by a preparatory command designated by the size of the unit, such as "**Squad, Platoon,**" or "**Company.**" For example, when drilling a squad the command to bring them to attention would be "**Squad, ATTENTION.**" Thereafter, they move only as ordered until given "**AT EASE,**" "**REST,**" "**FALL OUT,**" or they are dismissed.

9. When at a position of rest or while marching at route step or at ease and the command "**ATTENTION**" is given, the following applies:

   a. When given at parade rest, come to attention.

   b. When given at ease or rest, assume the position of parade rest on the preparatory command "**Squad.**" When "**ATTENTION**" is given go to that position.

   c. When given while marching at route step or at ease, get in step as soon as possible and continue to march at attention.

a. Front View.

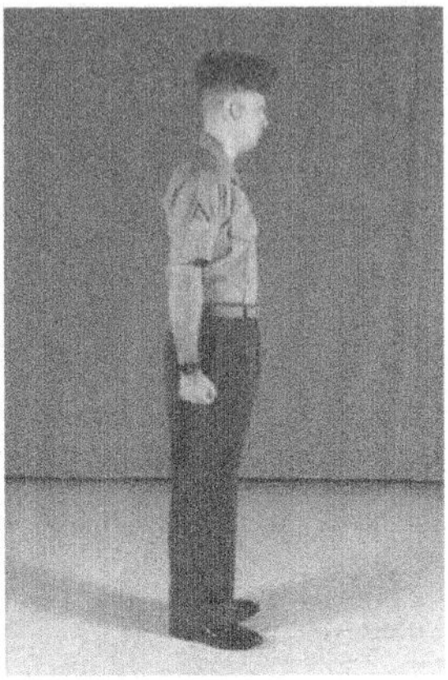

b. Side View.

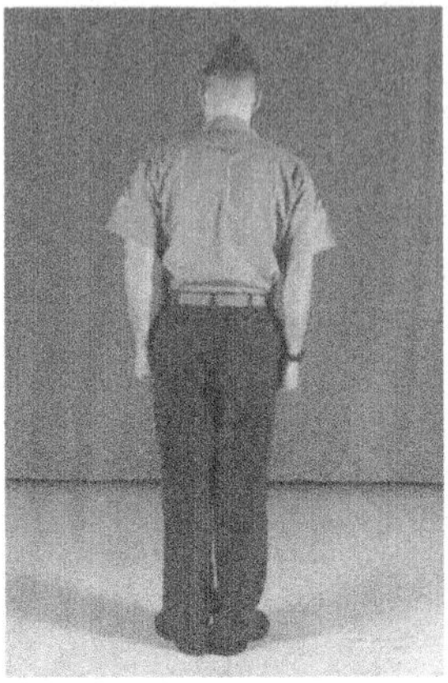

c. Rear View.

Figure 2-1.--Position of Attention.

2101. REST. There are four positions of rest for halted troops. They are parade rest, at ease, rest, and fall out. The purpose of rest is to give troops a rest from the position of attention. Parade rest, at ease, and rest, are one-count movements. Fall out is not a precision movement and has no counts. All are executed from the position of attention. The commands are "**Parade, REST**;" "**AT EASE**;" "**REST**;" and "**FALL OUT**."

1. Parade Rest. On the preparatory command "**Parade**," shift the weight of your body to the right leg without noticeable movement. On the command of execution "**REST**," and for the count of one, move the left foot twelve inches (measured from the inside of each heel) smartly to the left. The heels remain on line and the body weight rests equally on both legs. At the same time the left foot is moved, clasp the hands behind the back. The left hand is placed just below the belt and the right hand is placed inside the left. The thumb of the right hand lightly grasps the thumb of the left. All fingers are extended and joined with the palms to the rear. The elbows will be in line with the body. Silence and immobility are required. (See figure 2-2.) The only command you may receive while at parade rest is "**ATTENTION**."

2. At Ease. The command is "**AT EASE**." It is executed in one count. At the command, keep your right foot in place. You may move about, but must not talk. You may also be given this command when not in ranks. In this case, it means cease talking, but continue whatever you were doing before the command. When in ranks, the only command you may receive while at ease is "**ATTENTION**."

3. Rest. The command is "**REST**." It is executed in one count. At the command, you may move and talk; however, you must keep your right foot in place. The only command you may receive while at rest is "**ATTENTION**."

4. Fall Out. The command is "**FALL OUT**." At the command, leave your position in ranks, but remain nearby or proceed to a pre-designated area. When "**FALL IN**" is given, return to your place in ranks at the position of attention.

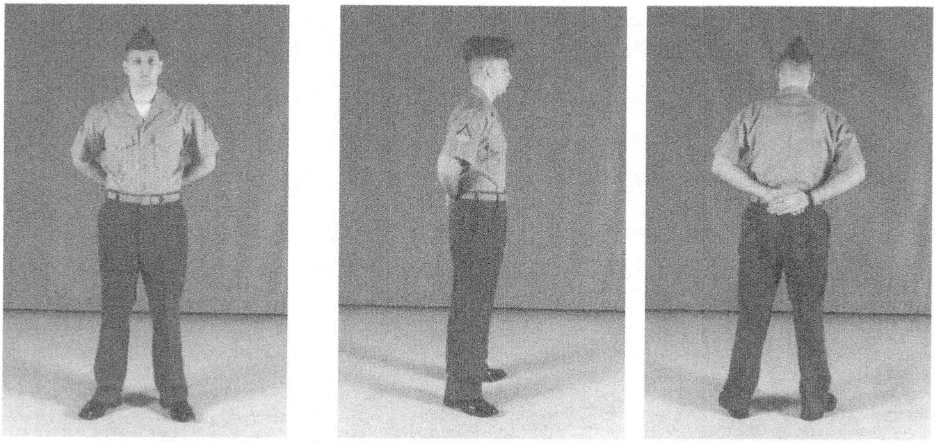

a. Front View. b. Side View. c. Rear View.

Figure 2-2.--Position of Parade Rest.

2102. <u>EYES RIGHT (LEFT)</u>. The purpose of eyes right (left) is to demonstrate military courtesy to reviewing officers and dignitaries during parades and ceremonies. It is executed in one count when halted at attention or marching at quick time. The command is **"Eyes, RIGHT (LEFT)."** The command to turn the head back to the position of attention is **"Ready, FRONT."**

1. When the command of execution **"RIGHT (LEFT)"** is given, turn your head smartly and look 45 degrees to the right (left) keeping your shoulders square to the front.

2. On the command of execution **"FRONT,"** turn your head and eyes smartly back to the front. During reviews at which the reviewing officer troops the line, ready front will not be given after eyes right. At such ceremonies, turn your head and eyes smartly toward the reviewing officer upon the command of execution **"RIGHT."** As he passes to the left, follow the reviewing officer with your head and eyes until you are looking directly to the front.

3. When marching, give the command, **"Eyes, RIGHT"** so that the command of execution is given as the right foot strikes the deck and the command **"Eyes, LEFT"** so that the command of execution is given as the left foot strikes the deck. Give the command **"Ready, FRONT"** from eyes right so that the command of execution is given as the left foot strikes the deck and **"Ready, FRONT"** from eyes left so that the command of execution is given as the right foot strikes the deck. For example:

LEFT	RIGHT	LEFT	RIGHT
	"Eyes,	(Step)	**RIGHT"**
"Ready,	(Step)	**FRONT"** (From **"Eyes, RIGHT"**)	
"Eyes,	(Step)	**LEFT"**	
	"Ready,	Step)	**FRONT"** (From **"Eyes, LEFT"**)

2103. FACING MOMENTS. The purpose of facing movements is to face a unit to the right, left or about. Facing movements are executed in two counts when halted at attention. The commands are **"Right, FACE;" "Left, FACE;"** and **"About, FACE."** Facing movements are executed in the cadence of quick time. While facing, your arms should not swing out from your sides, but remain at the position of attention.

1. **"Right, FACE"** is a two-count movement. (See figure 2-3.)

a. On count one, (see figure. 2-3b) at the command **"FACE,"** raise your left heel and right toe slightly. Turn to the right on your right heel and left toe. Keep your left leg straight but not stiff.

b. On count two (see figure 2-3c), place the left foot smartly beside the right and stand at attention.

2. **"Left, FACE"** is executed in the same manner described in 2103.1 above, substituting left for right and right for left.

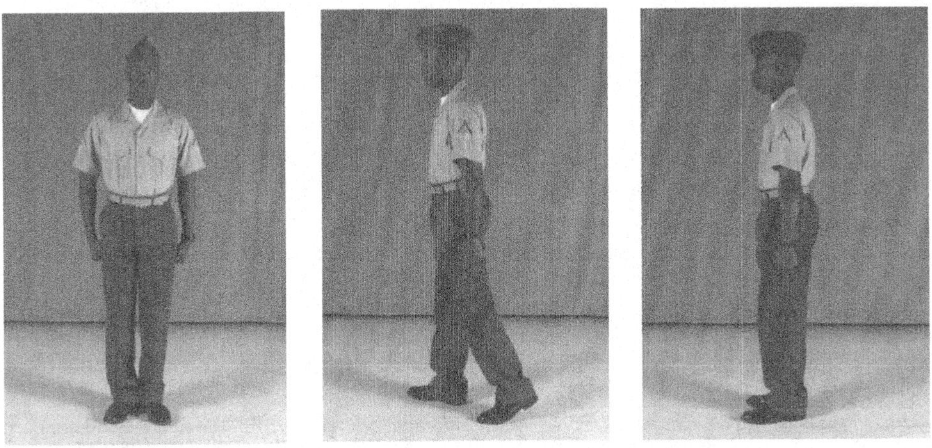

a. Start. b. Count One. c. Count Two.

Figure 2-3.--Right Face.

3. **"About, FACE"** is a two-count movement. (See figure 2-5.)

a. At the command **"About,"** shift your weight to your left leg without noticeable movement. (See figure 2-5a.)

b. On count one (see figure 2-5b) at the command **"FACE,"** place our right toe half a foot length behind and slightly to the left of your heel. Do not change the position of your left foot. Rest your weight evenly on the left heel and the ball of the right foot.

c. On count two (see figure 2-5c), turn smartly to the right until facing rear. The turn is made on the left heel and ball of the right foot. The knees remain straight, but not locked during the movement. Your thumbs will remain on the seams of your trousers. If properly executed you will be at the position of attention facing in the opposite direction.

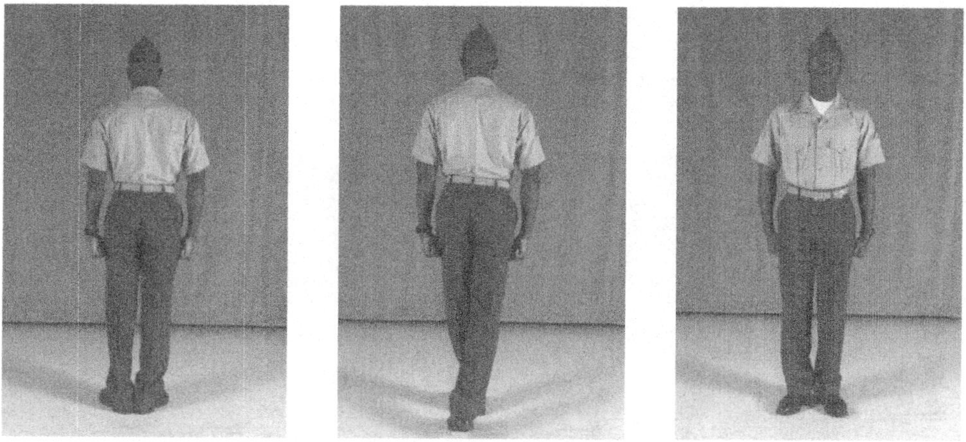

a. Count One. b. Count Two. c. Finish.

Figure 2-5.--About Face.

2104. HAND SALUTES

1. The purpose of the hand salute is to demonstrate mutual respect and courtesy between members of military organizations and to show respect to national colors. It is executed in one count when halted at attention, marching at quick time, or seated in a vehicle. The command is **"Hand, SALUTE."** To return to the position of attention the command is **"Ready, TWO."**

a. When **"SALUTE"** is given, raise your right hand smartly in the most direct manner until the tip of your forefinger touches the lower part of the headdress above and slightly right of your right eye. Your fingers should be extended straight and joined with the thumb along the forefinger. You should be able to see your entire palm when looking straight ahead. Your upper arm should be parallel with the deck with the elbow in line with the body and your forearm at a 45-degree angle. Your wrist and hand should be straight, a continuation of the line made by your forearm. At the same time, if not in ranks, turn your head and eyes toward the person or colors you are saluting.

b. At the command **"TWO,"** return to attention. Move your hand smartly in the most direct manner back to its normal position by your side.

c. To ensure simultaneous execution of the second movement of the hand salute when troops are in formation, the preparatory command **"Ready,"** will be used prior to the command of execution **"TWO."**

2. You may salute without command from attention, while walking, or while seated in a vehicle. When walking, it is not necessary to halt to salute. Keep walking, but at attention. The salute is rendered when the person or colors to be saluted is at a 6-pace distance, or at the nearest point of approach if it is apparent that the person or color is not going to approach within 6 paces. The salute will not be rendered if the person (color) to be saluted does not approach within 30 paces.

Hold the first position of the salute until the person (color) saluted has passed or the salute is returned, then execute the second movement of the hand salute.

3. When the command "**Present, ARMS**" is given, if not armed, you execute the hand salute on the command "**ARMS.**" Stay at that position until the command "**Ready, TWO**" or "**Order, ARMS**" is given.

Figure 2-6.--Hand Salute.

SECTION 2: STEPS AND MARCHING

2200. GENERAL

1. With the exception of right step, all steps and marching movements beginning from a halt start with the left foot.

2. The instructor indicates the proper rhythm by counting cadence.

3. To change direction on the march, the command of execution is given as the foot in the desired direction of the turn strikes the deck.

2201. QUICK TIME. The purpose of quick time is to march at 112 to 120 steps per minute taking 30-inch steps. It may be executed when halted at attention, marking time, marching forward at double time, or marching at half step. The commands are "**Forward, MARCH**" or "**Quick Time, MARCH**."

1. When halted, the command to march forward at quick time is "**Forward, MARCH**."

2. On "**Forward**," shift your weight to the right leg without noticeable movement.

3. On "**MARCH**," step off smartly, left foot first, and march straight ahead with 30-inch steps. Swing your arms easily in their natural arcs, 6 inches straight to the front and 3 inches to the rear. Do not exaggerate movements.

2202. DOUBLE TIME. The purpose of double time is to march at 180 steps per minute taking 36-inch steps. It may be executed when halted at attention, marking time, double timing in place, or marching forward at quick time. The command is "**Double Time, MARCH**." When given while marching, the command "**MARCH**" may be given as either foot strikes the deck.

1. When Halted

    a. At the command "**Double Time**," shift your weight to your right leg without noticeable movement.

    b. On "**MARCH**," step off on the left foot and begin an easy run, taking 180, 36-inch steps per minute. At the same time raise your forearms until parallel with the deck and make fists of your hands.

2. When Marching at Quick Time

    a. When the command "**MARCH**" is given, take one more 30-inch step.

    b. Step off in double time.

3. To Resume Quick Time, the Command is "**Quick Time, MARCH**"

    a. On "**MARCH**," take one more 36-inch step.

    b. Lower your arms and slow to quick time.

4. When Marching in Place

   a. Take one more step in mark time (or double time in place).

   b. Step off forward in double time.

2203. HALT. The purpose of halt is to stop the movement of a marching unit. It is executed in two counts when marching at quick time, and three counts when marching at double time. It may be executed when marching at quick time, half step, double time, side step, back step, mark time, and double time in place. The command of execution is "**HALT**," which is always preceded by a preparatory command designated by the size of the unit, such as "**Squad, Platoon**," or "**Company**." The command of execution may be given as either foot strikes the deck except for side step.

1. While Marching at Quick Time or Half Step

   a. Take one more step after "**HALT**."

   b. Bring the rear heel forward against the other.

2. While Marching at Double Time

   a. Take one more double time step after "**HALT**."

   b. Take another at quick time.

   c. Bring the rear heel forward against the other.

3. While Marching at Right (Left) Step

   a. The command "**HALT**" is given when the heels are together.

   b. Take one more side step after "**HALT**."

   c. Bring the heels together.

4. While Backward Marching

   a. Take one more back step after "**HALT**."

   b. Bring the forward heel back against the other.

5. While Marching at Mark Time

   a. Take one more step in place after the command "**HALT**."

   b. Bring the heels together.

6. While Marching at in Place Double Time

   a. Take one more in place double time step after the command "**HALT**."

   b. Take another at quick time.

   c. Bring the heels together.

2204. MARK TIME. The purpose of mark time is to march in place at quick time cadence. It may be executed when halted at attention, double timing in place, marching at half step and marching forward at quick time. While marching, the command may be given as either foot strikes the deck. The command is "**Mark Time, MARCH.**"

1. When Halted

a. On the command "**Mark Time,**" shift your weight to your right leg slightly without noticeable movement.

b. On the command "**MARCH,**" beginning with your left foot, then alternating, raise each foot so that the ball of the foot is approximately 2 inches and the heel approximately 4 inches from the deck at a cadence of quick time. At the same time, swing your arms naturally as in marching.

2. When Marching at Quick Time

a. On the command "**MARCH,**" take one more 30-inch step as if coming to a halt.

b. Bring your heels together. Begin marking time without loss of cadence with the opposite foot.

3. To Resume the March at Quick Time, The Command is "**Forward, MARCH**"

a. On the command "**MARCH,**" take one more step in place.

b. Step off with a 30-inch step.

4. When Marching at Double Time in Place

a. On the command "**MARCH,**" take one more step in place at double time.

b. Commence marking time.

2205. DOUBLE TIME IN PLACE. The purpose of double time in place is to march in place at double time cadence. It may be executed when halted at attention, marking time, or marching forward at double time. The command is "**In Place Double Time, MARCH.**"

1. When Halted

a. On the command "**In Place Double Time,**" shift your weight to your right leg without noticeable movement.

b. On the command "**MARCH,**" raise your arms for double time. Starting with the left foot, alternating and in place, raise your feet 6 inches above the deck at 180 steps per minute.

2. When Marching at Double Time

a. After "**MARCH,**" take two more double time steps.

b. Bring your feet together and begin double timing in place without loss of cadence.

3. The Command **"Double Time, MARCH"** Resumes the March at Double Time

a. At **"MARCH,"** take two more steps in place.

b. Step off with 180, 36-inch steps per minute.

4. When Marking Time

a. On the command **"MARCH,"** take one more step in mark time.

b. Commence marching at double time in place.

2206. HALF STEP. The purpose of half step is to march forward at quick time taking 15-inch steps. It may be executed when halted at attention or marching forward at quick time. While marching it may be given as either foot strikes the deck. The command is **"Half Step, MARCH."**

1. At the Halt

a. On the command **"Half Step,"** shift your weight to your right leg without noticeable movement.

b. On the command **"MARCH,"** step off forward in quick time cadence with 15-inch steps. The balls of your feet should strike the deck before your heels. At the same time begin swinging your arms 6 inches to the front and 3 inches to the rear.

2. Marching in Quick Time

a. On the command **"MARCH,"** take one more 30-inch step.

b. Begin the 15-inch steps as explained above. (See paragraph 2206.1.)

3. To resume quick time from half step, the command is **"Forward, MARCH."** It may be given as either foot strikes the deck.

a. On **"MARCH,"** take one more half step.

b. Step off at quick time.

2207. SIDE STEP. The purpose of side step is to move the unit a short distance to the right or left. There are no counts, however there is a cadence of 1-2.

1. The command **"Right (Left) Step, MARCH"** is given only when you are at a halt.

2. At the command **"MARCH,"** move your right foot 12 inches to the right, then place your left foot smartly beside your right. Repeat this movement at quick time. Keep your legs straight, but not stiff. Hold your arms at your sides as in the position of attention.

2208. BACK STEP. The purpose of back step is to march the unit backwards for a short distance.

1. The command **"Backward, MARCH"** is given only when you are at a halt.

2. At the command **"MARCH,"** step off to the rear with your left foot and take 15-inch steps at quick time. At the same time swing your arms 6 inches to the front and 3 inches to the rear.

2209. TO FACE TO THE RIGHT (LEFT) IN MARCHING

1. This is an important part of the following movements: column right (left), by the right (left) flank, close, take interval, and extend. For instructional purposes, the command is **"By The Right (Left) Flank, MARCH."**

2. From a Halt, on the Command **"MARCH"**

a. For right flank, turn 90 degrees to the right by pivoting on the ball of the right foot and (using a cross over step) stepping of with the left foot 30 inches in the new direction of march.

b. For left flank, turn 90 degrees to the left by pivoting on the ball of the right foot and stepping off 30 inches with the left foot in the new direction of march.

3. While marching, the command of execution is given as the foot of the desired direction of movement strikes the deck.

a. On **"MARCH,"** without losing cadence, take one more step in the original direction.

b. Pivot to the right (left) and step off in the new direction. If commanded and executed properly, the pivot will be on the foot away from the desired direction of movement, and the first step in the new direction will be with the other foot. The pivot and step are done together in one count. When marching to the flank the only other movement that can be given is a flanking movement in the opposite direction that will bring the individual or unit back to the original front.

2210. TO FACE TO THE REAR WHILE MARCHING. The purpose of this movement is to march the unit to the rear for a short distance.

1. The command is **"To the Rear, MARCH."** It may be executed when halted or while marching at either quick time or double time. When marching the command of execution **"MARCH"** is given when the right foot strikes the deck. When marching to the rear, the only movement that can be given is **"To the Rear, MARCH"** in order to bring the individual or unit back to its' original front.

2. To March to the Rear From a Halt

a. The command is **"To the Rear, MARCH."**

b. On **"MARCH,"** step forward with the left foot.

c. Turn about to the right on the balls of both feet and immediately step out to the rear with the left foot. This is done without loss of cadence. When turning, the feet should be about 15 inches apart.

3. When at Quick Time

a. On the command "**MARCH,**" take one more step.

b. Turn about to the right on the balls of both feet and immediately step off to the rear with your left foot. This is done without loss of cadence. When turning, the feet should be about 15 inches apart.

4. When at Double Time

a. On "**MARCH,**" follow the sequence of steps and actions below.

As this foot strikes the deck	Take the following action
RIGHT	"**MARCH**" is given.
LEFT	36-inch step.
RIGHT	36-inch step. Then in place double time.
LEFT	Pivot 90 degrees to the right
RIGHT	Heels together.
LEFT	Pivot 90 degrees to the right.
RIGHT	Heels together.
LEFT	36-inch step in the new direction.

2211. TO MARCH TO THE FLANK. The purpose of this movement is to march the entire unit to the right (left) for a short distance. It may be executed when halted or while marching at either quick time or double time. The command is "**By the Right (Left) Flank, MARCH.**" When marching the command of execution is given as the foot in the direction of the movement strikes the deck.

1. From a Halt

a. For right flank, turn 90 degrees to the right by pivoting on the ball of the right foot and (using a cross over step) stepping off with the left foot 30 inches in the new direction of march.

b. For left flank, turn 90 degrees to the left by pivoting on the ball of the right foot and stepping off 30 inches with the left foot in the new direction of march.

2. While at Quick Time

a. On "**MARCH,**" take one more step.

b. Turn to the right (left) on the ball of the left (right) foot. At the same time, step off 30 inches in the new direction with the right (left) foot.

3. While Double Timing

a. On "**MARCH,**" use the following sequence and actions.

By the Left Flank

Step	Action
LEFT	"**MARCH**" is given.
RIGHT	36-inch step.
LEFT	in place double timing.
RIGHT	Pivot 90 degrees to the left.
LEFT	36-inch step in new direction.

By the Right Flank

Step	Action
RIGHT	"**MARCH**" is given.
LEFT	36-inch step in place double timing.
LEFT	Pivot 90 degrees to the right.
RIGHT	36-inch step in new direction.

2212. TO CHANGE STEP. The purpose of this movement is to change the cadence count without changing the rhythm of the cadence.

1. The command is **"Change Step, MARCH."** It may be given while marching at quick or double time, marking time, or double timing in place. The command of execution is given as the right foot strikes the deck.

2. While Marching at Quick Time or Double Time

a. On **"MARCH,"** take one more step, 30 or 36 inches, as appropriate.

b. As your right foot comes forward to the next step, place the toe near the left heel and step out again with the left foot. This changes the cadence count, but not the rhythm.

3. While Marking Time

a. On **"MARCH,"** lift and lower the left foot twice in succession.

b. The second time it touches the deck, raise the right foot and continue marking time.

4. While Double Timing in Place

a. On **"MARCH,"** hop twice on the left foot.

b. Continue double timing in place.

2213. TO MARCH AT EASE. The purpose of this movement is to give troops a rest from marching at attention or when keeping in step becomes difficult due to rough terrain.

1. The command is **"At Ease, MARCH."** It is given as either foot strikes the deck and only while marching at quick time.

2. After **"MARCH,"** you are no longer required to march in cadence, but must keep interval and distance. Do not talk.

2214. TO MARCH AT ROUTE STEP

1. The command is **"Route Step, MARCH."** It is commanded as either foot strikes the deck and only while marching at quick time.

2. After **"MARCH,"** you are no longer required to march in cadence, but must keep interval and distance. You may talk, but in a low voice.

2215. TO RESUME MARCHING AT QUICK TIME FROM ROUTE STEP OR AT EASE

1. The command is **"Squad (Platoon, Company, Detail, etc.), ATTENTION."**

2. After commanding "**ATTENTION,**" the unit leader will count cadence until all troops are in step. Pick up the step as soon as possible and continue marching at attention.

3. Commands that involve precision in execution will not be given until the command to resume marching at quick time has been given and all members of the formation have picked up the step and cadence.

MARINE CORPS DRILL AND CEREMONIES MANUAL

PART I: DRILL

CHAPTER 3

MANUAL OF ARMS WITH THE M16 RIFLE

SECTION 3: MANUAL OF ARMS MOVEMENTS WITH THE M203

FIGURE

MARINE CORPS DRILL AND CEREMONIES MANUAL

PART I: DRILL

CHAPTER 3

MANUAL OF ARMS WITH THE M16 RIFLE

3000. GENERAL. This Chapter covers the manual of arms movements for the M16 rifle with both the parade and loosened sling, and the M16 with the M203 grenade launcher attached. Whenever executing any manual of arms movements ensure that the weapon has been cleared and that all weapons handling procedures are followed.

MARINE CORPS DRILL AND CEREMONIES MANUAL

CHAPTER 3

MANUAL OF ARMS WITH THE M16 RIFLE

SECTION 1: MANUAL OF ARMS WITH A PARADE SLING

3100. GENERAL

1. Prior to commencing the manual of arms, the magazine is removed, and the sling is drawn tight and positioned on the left (selector switch) side of the rifle. This configuration is called a parade sling and facilitates execution of the manual of arms. Figure 3-1 identifies the drill nomenclature for the M16 rifle with sling.

2. All movements of the manual of arms are executed smartly and with snap.

3. On the command "**FALL IN**," if armed with a rifle it will be positioned at order arms.

4. Facings, alignments, and short distance marching movements are executed from order arms. Side step, back step, open and close ranks, and close and extend are short distance movements. Forward march may be given from order arms to march units forward for a short distance. When these movements are commanded while at order arms, it is necessary to come automatically to trail arms and step off simultaneously on the command of execution for the movement. The rifle is returned to order arms upon halting.

5. Before a command for any marching movement (other than the executions stated in paragraph 3100.3) is given to armed troops, they are faced in the direction of march and their weapons brought to right (left) shoulder, port, or sling arms by the appropriate command. After a marching movement has been completed and it is desired to execute a facing movement, the command to order or unsling arms is given, followed by the command for the facing movement.

6. When at a position other than sling arms, the troops must come to port arms for double time. When the troops are in formation, the commander gives the appropriate commands.

7. Generally, the M16 rifle should be carried at sling arms except for ceremonies and drill periods.

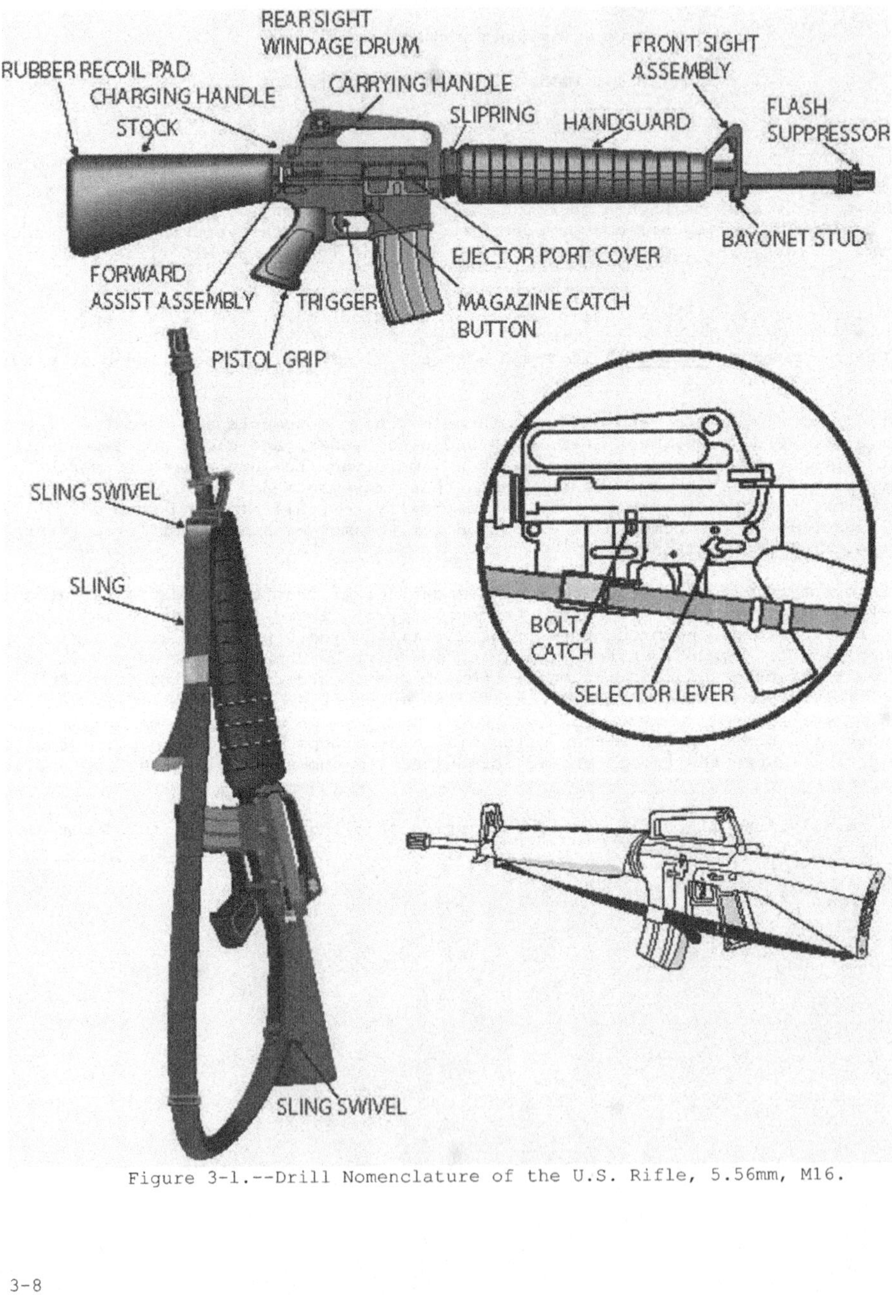

Figure 3-1.--Drill Nomenclature of the U.S. Rifle, 5.56mm, M16.

3101. <u>ORDER ARMS</u>

1. The basic position of the rifle manual is order arms. (See figure 3-2). It is the position assumed by an individual halted at attention with the rifle. It is assumed on the command **"ATTENTION"** from any of the rest positions except fall out. Order arms is also assumed on the command **"<u>FALL IN</u>"** and on the command **"Order, ARMS,"** from any position in the manual except inspection arms and sling arms.

2. Properly executed, the butt of the rifle rests on the deck, and the stock rests along the outer edge of the right shoe. The pistol grip is to the front and the barrel is in a near vertical position. The rifle is grasped by the right hand at the junction of the front sight assembly and the barrel. The barrel rests in the "V" formed by the thumb and forefinger. The fingers are extended and joined and placed on line with the barrel. The thumb is placed along the trouser seam and the entire right arm is behind the rifle. This may cause a slight bend to the right arm of shorter individuals. Persons, too tall to reach the junction of the barrel and front sight assembly, may slide their hand up the barrel so that their posture remains erect.

a. Front. b. Right Side.

Figure 3-2.--Order Arms.

3102. REST POSITIONS

1. The purpose of rest positions with the rifle is the same as rest positions without arms. They are commanded and executed as without arms with the following exceptions and additions.

2. On the command "**Parade, REST,**" the left foot is moved 12 inches to the left. The rifle butt is kept against the right foot. The grasp of the right hand is moved up the barrel to just below the flash suppresser with the fingers joined and curled with the forefinger touching the thumb. The right arm is straightened directly to the front so that the muzzle points forward and up. (See figure 3-3a.) At the same time the left foot is moved, the left hand is placed behind the back just below the belt line with fingers and thumb extended and joined and the palm facing the rear. (See figure 3-3b.)

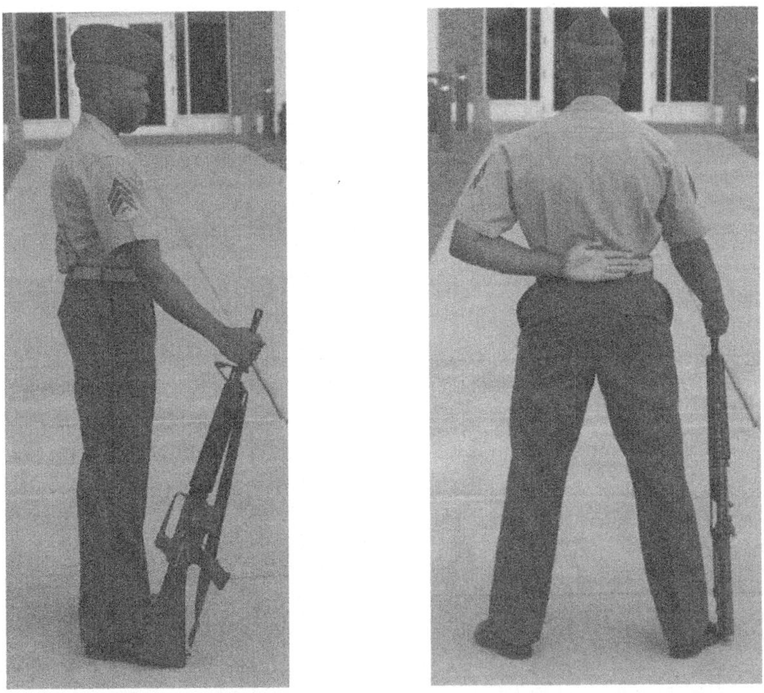

a. Side View. b. Rear View.

Figure 3-3.--Parade Rest.

3. On the command "**AT EASE**" or "**REST,**" the right foot remains in place and the rifle is held as in parade rest with the right arm relaxed slightly.

3103. TRAIL ARMS

1. The purpose of trail arms is to raise the rifle off the deck when facing, aligning or moving a short distance. It is executed in one count simultaneously with the first count of the movement commanded, and upon that movement's command of execution. It may be executed only when halted at order arms. The rifle is returned to order arms quietly and without command upon completion of the movements.

2. For training purposes, trail arms may be executed on command. The command is "**Trail, ARMS.**" At the command of execution "**ARMS,**" the rifle is raised vertically 3 inches off the deck. The wrist is kept straight with the thumb along the seam of the trousers. (See figure 3-4.)

3. At the command "**Order, ARMS,**" the rifle is quietly lowered to the position of order arms.

a. Front b. Right Side

Figure 3-4.--Trail Arms.

3104. PORT ARMS FROM ORDER ARMS

1. The purpose of this movement is to bring the rifle to a two-handed carry position. This is a two-count movement from order arms. It is executed when halted at order arms. The command is "**Port, ARMS**."

2. At the command of execution, and for the count of one, slide the right hand up and grasp the barrel near the flash suppressor. The fingers are joined and wrapped around the barrel with the thumb wrapped around the inboard portion. Without loss of motion, raise and carry the rifle diagonally across the front of the body until the right hand is level with and slightly to the left of the face. The right wrist is on the outboard portion of the front sight assembly. The elbow is held down without strain and nearly touches the handguard. The barrel is up and bisecting the angle formed by the neck and left shoulder. The pistol grip is to the left. The butt is in front of the right hip. At the same time, smartly grasp the handguard with your left hand, just above the slipring. The sling is included in the grasp. The fingers are joined and grasping the rifle. The little finger is in line with the slipring with the thumb on the inboard side of the handguard. The left wrist and forearm are straight. The elbow is held in against the body. (See figure 3-5b.)

3. On the second count, release the grasp of the right hand and smartly re-grasp the small of the stock. The fingers are joined and wrapped around the small of the stock with the thumb wrapped around the inboard portion. The right wrist and forearm are straight and parallel to the deck. The elbow is held into the side and the upper arm is in line with the back. The rifle is about 4 inches from the body. (See figure 3-5c.)

a. Start. b. Count One. c. Count Two.

Figure 3-5.--Movements from Order Arms to Port Arms.

3105. MOVEMENTS FROM PORT ARMS

1. The purpose of these movements is to move the rifle from port arms to order arms, left shoulder arms, right shoulder arms and present arms.

2. Order arms is a three-count movement. It is executed only when halted at port arms. The command is "**Order, ARMS.**"

a. At the command of execution, and for the count of one, the right hand is moved from the small of the stock and smartly re-grasps the barrel. The palm of the right hand is to the rear. The fingers are joined and wrapped around the barrel with the thumb wrapped around the inboard portion. The little finger is just above the bayonet stud. The right wrist is on the outboard portion of the front sight assembly. The elbow is held down without strain and nearly touches the handguard. (See figure 3-6b.)

b. On the second count, lower the rifle initially with the left hand while changing the grasp of the right hand to the junction of the barrel and the front sight assembly as in the position of order arms. Without loss of motion, release the grasp of the left hand from the handguard and, with your right hand, carry the weapon to your right side until the butt is 3 inches from the deck. The barrel is in a vertical position and the pistol grip is to the front. At the same time guide the weapon with the left hand until the right thumb is on the trouser seam. The fingers of the left hand are extended and joined and touch the rifle near the flash suppresser. The palm of the left hand is toward the rear. The left wrist and forearm are straight and the left elbow is in against the body. (See figure 3-6c.)

c. On the third count, return the left hand to the left side at the position of attention. At the same time, gently lower the rifle to the deck with the right hand. (See figure 3-6d.)

a. Start. b. Count One. c. Count Two. d. Count Three.

Figure 3-6.--Movements from Port Arms to Order Arms.

3. Left shoulder arms from port arms is a two-count movement. It may be executed while halted at attention or while marching. The command is "**Left Shoulder, ARMS.**"

a. At the command of execution, and for the count of one, release the grasp of the left hand from the handguard and, with the right hand, carry the rifle to the left side rotating it a quarter turn counterclockwise. Place it on the left shoulder. The elbow is held down without strain. At the same time, grasp the butt of the rifle with the left hand in the same manner as for right shoulder arms. (See figure 3-7b.)

b. On the second count, move your right hand back to the right side at the position of attention. The rifle is held at a 60-degree angle from the deck. (See figure 3-7c.)

a. Start. b. Count One. c. Count Two.

Figure 3-7.--Movements from Port Arms to Left Shoulder Arms.

4. Right shoulder arms from port arms is a three-count movement. It may be executed while halted at attention or while marching. The command is "**Right Shoulder, ARMS.**"

a. On the command of execution, and for the count of one, move the right hand from the small of the stock and grasp the butt placing the heel of the butt between the first two fingers. The heel will be visible between the index and middle fingers. (See figure 3-8b.)

b. On the second count, release the grasp of the left hand from the handguard and carry the rifle to the right side. The left hand remains in contact with the rifle in order to assist in the movement. At the same time, with the right hand, rotate the rifle a quarter turn counterclockwise and place the rifle into the right shoulder. Slide the left hand to the junction of the stock and receiver just below the charging handle. Guide the rifle into the shoulder with the left hand. The thumb and fingers are extended and joined with the palm turned toward the body. The first joint of the left forefinger touches the rear of the receiver. The left wrist and forearm are straight. The left elbow is held against the body. The grasp of the right hand is unchanged. The right wrist and forearm are straight and parallel to the deck. The elbow is held into the side with the upper arm in line with the back. (See figure 3-8c.)

c. On the third count, move your left hand back to the left side at the position of attention. The rifle is held at a 60-degree angle from the deck. (See figure 3-8d.)

a. Start. b. Count One. c. Count Two. d. Count Three.

Figure 3-8.--Movements from Port Arms to Right Shoulder Arms.

5. Present arms from port arms is a one-count movement. It is only executed while halted at attention. The movement may be executed without command, for a sentry on post, or at the command of "**Present, ARMS.**" To execute the movement the rifle is rotated clockwise with the right hand, the muzzle moves to the right, and the rifle is re-grasped above the slipring with the left hand. The sling is included in the grasp. (See figure 3-9.)

a. Start. b. Present Arms.

Figure 3-9.--Movements from Port Arms to Present Arms.

3106. RIGHT SHOULDER ARMS FROM ORDER ARMS

1. The purpose of this movement it to bring the rifle to a position on the right shoulder. This is a four-count movement from order arms. It is executed when halted at order arms. The command is "**Right Shoulder, ARMS**."

2. On the command of execution, and for the count of one, grasp the barrel of the weapon with the right hand. Without loss of motion, raise and carry the rifle diagonally across the front of the body until the right hand is level with and slightly to the left of the face. The right wrist and forearm are straight. The right elbow is held down without strain. The barrel is up and bisecting the angle formed by the neck and left shoulder. At the same time, grasp the handguard with the left hand (as in count one for port arms). (See figure 3-10b.)

3. On the second count, release the barrel with the right hand and grasp the butt placing the heel of the butt between the first two fingers. The heel will be visible between the index and middle fingers. (See figure 3-10c.)

4. On the third count, release the grasp of the left hand from the handguard and carry the rifle to the right side. The left hand remains in contact with the rifle in order to assist in the movement. At the same time, with the right hand, rotate the rifle a quarter turn counterclockwise and place the rifle into the right shoulder. Slide the left hand to the junction of the stock and receiver just below the charging handle. Guide the rifle into the shoulder with the left hand. The thumb and fingers are extended and joined with the palm turned toward the body. The first joint of the left forefinger touches the rear of the receiver. The left wrist and forearm are straight. The left elbow is held against the body. The grasp of the right hand is unchanged. The right wrist and forearm are straight and parallel to the deck. The elbow is held into the side with the upper arm in line with the back. (See figure 3-10d.)

5. On the fourth count, move your left hand back to the left side at the position of attention. The rifle is held at a 60-degree angle from the deck. (See figure 3-10e.)

a. Start. b. Count One.

Figure 3-10.--Movements from Order Arms to Right Shoulder Arms.

c. Count Two. d. Count Three. e. Count Four.

Figure 3-10.--Movements from Order Arms to Right Shoulder Arms--Continued.

3107. MOVEMENTS FROM RIGHT SHOULDER ARMS

1. The purpose of these movements is to move the rifle from right shoulder arms to port arms, order arms, left shoulder arms or present arms.

2. Port arms from right shoulder arms is a two-count movement. It may be executed while halted at attention or when marching. The command is "**Port, ARMS.**"

a. At the command of execution, and for the count of one, pull the rifle butt back quickly with the right hand so the rifle comes off the right shoulder. At the same time, with the right hand rotate the rifle a quarter turn clockwise so that the pistol grip is to the left. Allow the rifle to fall diagonally across the front of the body. Bring the left hand up and smartly grasp the handguard with the sling included in the grasp. The fingers are joined grasping the handguard. The little finger is in line with the slipring but not touching it. The thumb is on the inboard side. The left wrist and forearm are straight. The elbow is held in against the body. The barrel is up; bisecting the angle formed by your neck and left shoulder. The butt is in front of the right hip. The grasp of the right hand has not changed. The right arm is nearly extended with the elbow held against the body. (See figure 3-11b.)

b. On the second count, release the grasp of the right hand from the butt and smartly re-grasp the small of the stock. The fingers are joined and wrapped around the small of the stock with the thumb wrapped around the inboard portion. The right wrist and forearm are straight and parallel to the deck. The elbow is held into the side and the upper arm is in line with the back. The rifle is about 4 inches from the body and now in the position of port arms. (See figure 3-11c.)

a. Start. b. Count One. c. Count Two.

Figure 3-11.--Movements from Right Shoulder Arms to Port Arms.

3. Order arms from right shoulder arms is a four-count movement. Order arms may be executed only when halted at attention. The command is "**Order, ARMS**."

a. At the command of execution, and for the count of one, pull the rifle butt back quickly with the right hand so the rifle comes off the right shoulder. At the same time, with the right hand rotate the rifle a quarter turn clockwise so that the pistol grip is to the left. Allow the rifle to fall diagonally across the front of the body. Bring the left hand up and smartly grasp the handguard with the sling included in the grasp. The fingers are joined grasping the handguard. The little finger is in line with the slipring but not touching it. The thumb is on the inboard side. The left wrist and forearm are straight. The elbow is held in against the body. The barrel is up bisecting the angle formed by your neck and left shoulder. The butt is in front of the right hip. The grasp of the right hand has not changed. The right arm is nearly extended with the elbow held against the body. (See figure 3-12b.)

b. On the second count, release the grasp of the right hand from the butt and smartly re-grasp the barrel. The palm of the right hand is to the rear. The fingers are joined and wrapped around the barrel with the thumb wrapped around the inboard portion. The little finger is just above the bayonet stud and the right wrist in on the outboard portion of the front sight assembly. (See figure 3-12c.)

c. On the third count, lower the rifle initially with the left hand while changing the grasp of the right hand to the junction of the barrel and the front sight assembly as in the position of order arms. Without loss of motion, release the grasp of the left hand from the handguard and with your right hand, carry the weapon to your right side until the butt is 3 inches from the deck. At the same time, guide the weapon with the left hand until the right thumb is on the trouser seam. The fingers of the left hand are extended and joined and touching the rifle, near the flash suppresser. The palm of the left hand is toward the rear. The wrist and forearm are straight and the left elbow is in against the body. (See figure 3-12d.)

d. On the fourth count, return the left hand to the left side at the position of attention. At the same time, gently lower the rifle to the deck with the right hand. (See figure 3-12e.)

a. Start.

b. Count One.

c. Count Two.

d. Count Three.

e. Count Four.

Figure 3-12.--Right Shoulder Arms to Order Arms.

4. Left shoulder arms from right shoulder arms is a four-count movement. It may be executed while halted at attention or while marching. The command is "**Left Shoulder, ARMS**." The first two counts move the rifle to port arms as described in paragraph 3107.2. The second two counts carry the rifle into the left shoulder as described in paragraph 3105.3.

5. Present arms from right shoulder arms is a three-count movement. The first two counts move the weapon to port arms as described in paragraph 3107.2. For the third count rotate the weapon to present arms as described in paragraph 3105.5.

3108. LEFT SHOULDER ARMS FROM ORDER ARMS

1. The purpose of this movement it to bring the rifle to a position on the left shoulder. This is a four-count movement from order arms. It is executed when halted at order arms. The command is "**Left Shoulder, ARMS**."

2. At the command "**ARMS**," the rifle is brought to port arms on the first two counts. (See figures 3-13b and c.)

3. On the third count, release the grasp of the left hand from the handguard and, with the right hand, carry the rifle to the left side rotating it a quarter turn counterclockwise. Place it on the left shoulder. The elbow is held down without strain. At the same time, grasp the butt of the rifle with the left hand in the same manner as for right shoulder arms. (See figure 3-13d.)

4. On the fourth count, move your right hand back to the right side at the position of attention. The rifle is held at a 60-degree angle from the deck. (See figure 3-13e.)

a. Start. b. Count One.

Figure 3-13.--Movements from Order Arms to Left Shoulder Arms.

c. Count Two. d. Count Three. e. Count Four.

Figure 3-13.--Movements from Order Arms to Left Shoulder Arms--Continued.

3109. MOVEMENTS FROM LEFT SHOULDER ARMS

1. The purpose of these movements is to move the rifle from left shoulder arms to port arms, order arms, right shoulder arms or present arms.

2. Port arms is a two-count movement. It is executed when halted or marching at left shoulder arms. The command is "**Port, ARMS**."

a. On the command of execution, and for the count of one, the right hand is moved up across the body and grasps the rifle at the stock below the charging handle. (See figure 3-14b.)

b. On the second count, the grasp of the left hand is released, and the rifle is brought from the shoulder diagonally across the body with the right hand and re-grasped at the handguard just above the slipring with the left hand as in the position of port arms. (See figure 3-14c.)

3. Order arms from left shoulder arms is a five-count movement. Order arms may be executed only when halted at attention. The command is "**Order, ARMS**." For the first two counts, move the rifle to port arms as described in figures 13d and e. For the last three counts, the rifle is brought to order arms as described in figures 3-12c, d, and e.

4. Right shoulder arms from left shoulder arms is a five-count movement. The movement may be executed while halted at attention or while marching; the command is "**Right Shoulder, ARMS**." For the first two counts of the movement the rifle is moved to port arms. (See figure 3-14.) The next three counts move the rifle to right shoulder arms as described in paragraph 3105.4.

a. Start. b. Count One. c. Count Two.

Figure 3-14.--Movements from Left Shoulder Arms to Port Arms.

5. Present arms from left shoulder arms is a three-count movement. The movement may be executed only while halted at attention. The movement may be executed without command by a sentry on post or on the command **"Present, ARMS."** The first two counts bring the rifle to port arms described in paragraph 3109.2. For the third count rotate the weapon to present arms as described in paragraph 3105.5.

3110. PRESENT ARMS FROM ORDER ARMS

1. The purpose of this movement is to render a salute to persons or colors when in formation or when posted as a sentry. Order arms to present arms is a two-count movement and is executed when halted at order or port arms. The command is **"Present, ARMS."**

a. At the command of execution, and for the count of one, slide the right hand up and grasp the barrel near the flash suppresser. The fingers are joined and wrapped around the barrel. Without loss of motion, raise and carry the rifle to a vertical position centered on the body. The pistol grip is to the front. The wrist is on the right side of the front sight assembly. The elbow is held down without strain. At the same time, smartly grasp the rifle at the handguard with the left hand just above the slipring. The sling is included in the grasp. The fingers are joined and wrapped around the handguard. The little finger is on line with the slipring with the thumb on the inboard side of the handguard. The left wrist and forearm are straight and parallel to the deck. The elbow is held into the side with the upper arm on line with the back. (See figure 3-15b.)

b. On the second count, release the grasp of the right hand and re-grasp the small of the stock. The charging handle rests on the thumb of the right hand. The fingers are extended and joined diagonally across the small of the stock. The right wrist and forearm remain straight. The elbow is slightly bent and held against the body. The left thumb is 4 inches from the body. (see figure 3-15c.)

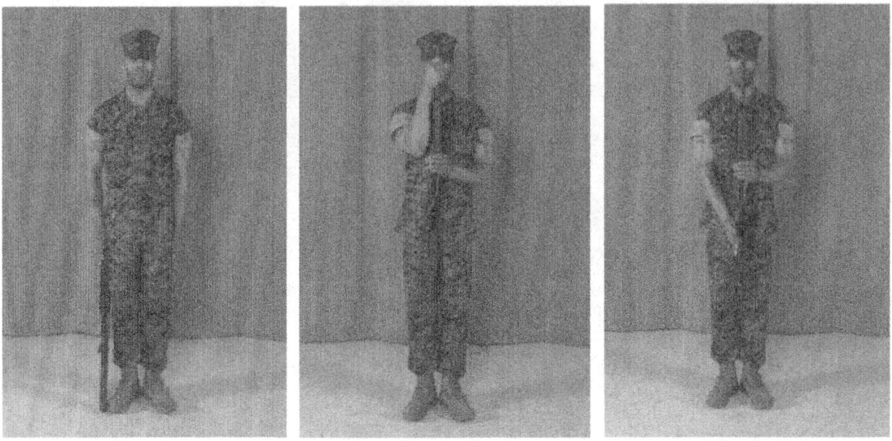

a. Start. b. Count One. c. Count Two.

Figure 3-15.--Movements from Order Arms to Present Arms.

3111. MOVEMENTS FROM PRESENT ARMS

1. The purpose of these movements is to terminate present arms and to move the rifle from present arms to order arms, port arms, and left or right shoulder arms. These movements are only executed when halted at present arms.

2. Order arms from present arms is a three-count movement. The command is **"Order, ARMS."**

a. At the command of execution, and for the count of one, release the grasp of the right hand from the small of the stock and re-grasp the barrel. The palm of the right hand is to the left. The fingers are joined and wrapped around the barrel. The little finger is just above the bayonet stud. The wrist is on the right side of the front sight assembly. The elbow is held down without strain. (See figure 3-16b.)

b. On the second count, lower the rifle initially with the left hand while changing the grasp of the right hand to the junction of the barrel and the front sight assembly as in the position of order arms. Without loss of motion, release the grasp of the left hand from the handguard and, with the right hand, carry the weapon to the right side until the butt is 3 inches from the deck. The barrel is in a vertical position and the pistol grip is to the front. At the same time, guide the weapon into the right side with the left hand. The fingers of the left hand are extended and joined with the thumb along the hand. The tips of the forefinger and middle finger are touching metal at a point near the flash suppresser. The palm is towards the rear. The left wrist and forearm are straight. Guide the weapon into the right side so that the thumb of the right hand will be on the trouser seam and push back on the barrel until the toe of the stock is on line with the toe of the right shoe. The entire right arm is behind the rifle. (See figure 3-16c.)

c. On the third count, quietly lower the rifle to the deck with the right hand so that the toe of the rifle is on line with the toe of the right shoe, and the barrel is in a near vertical position; at the same time smartly and in the most direct manner return the left hand to the side as in the position of attention. The thumb of the right hand remains along the trouser seam and the right arm remains behind the rifle. (See figure 3-16d.)

a. Start.

b. Count One.

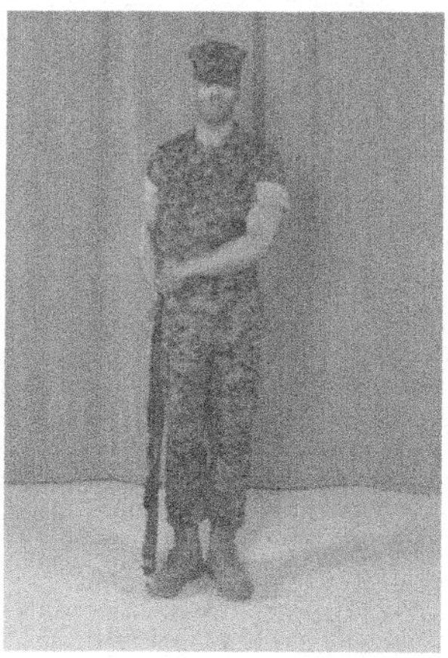

c. Count Two.

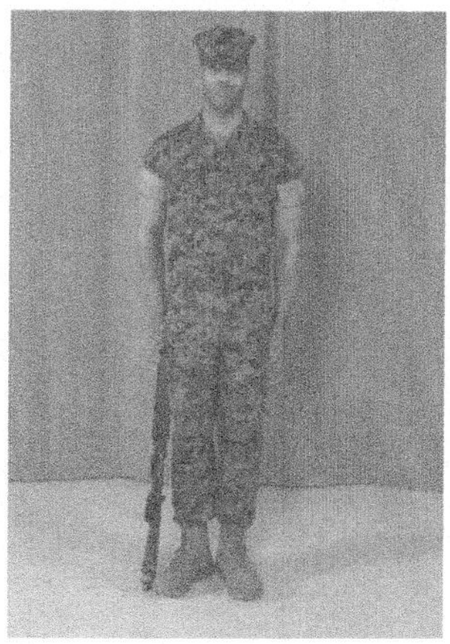

d. Count Three.

Figure 3-16.--Movements from Present Arms to Order Arms.

3. When a member of a color guard or when posted as a sentry, port arms, left or right shoulder arms may be executed from present arms.

a. Port arms is executed either without command or on the command of "**Port, ARMS.**" It is executed in one count by rotating the rifle counterclockwise with the right hand as the grasp of the left hand is adjusted on the handguard to that of port arms.

b. Left shoulder arms from present arms is a three-count movement. The movement is executed without command or on the command(s) of "**Left Shoulder, ARMS;**" "**Carry, COLORS;**" or "**Shoulder, ARMS.**" The first count is to rotate the rifle to port arms as described in paragraph 3111.3a. The next two counts are to move the rifle from port arms to the left shoulder as described in paragraph 3105.3.

c. Right shoulder arms from present arms is a four-count movement. The movement is executed without command or on the command(s) of "**Right Shoulder, ARMS;**" "**Carry, COLORS;**" or "**Shoulder, ARMS.**" The first count is to rotate the rifle to port arms as described in paragraph 3111.3a. The next three counts are to move the rifle from port arms to the right shoulder as described in paragraph 3105.4.

3112. RIFLE SALUTE

1. The purpose of this movement is to render a salute to persons or colors when not in formation nor posted as a sentry. It is a one-count movement executed when halted at order arms, trail arms, or when halted or marching at right or left shoulder arms. The command is "**Rifle, SALUTE**" and to terminate the salute the command is "**Ready, TWO.**"

2. When at order arms, the left arm moves smartly across the body with the forearm and wrist straight, fingers extended and joined and palm down. The first joint of the forefinger touches the flash suppresser. When not in ranks, the head turns toward the person or colors saluted. (See figure 3-17a.) On "**Ready, TWO**" resume the position of attention.

3. When at trail arms, the movements are identical with those for saluting at order arms, except that the rifle is held in the trail arms position. (See figure 3-17b.)

4. At right (left) shoulder arms, the left (right) arm moves across the body, fingers extended and joined, and palm down. The first joint of the forefinger touches the rear of the receiver just below the charging handle, and the forearm is held parallel to the deck. (See figures 3-17c and d.) When not in ranks, the head and eyes turn toward the person or colors saluted. On "**Ready, TWO**" the position of attention is resumed.

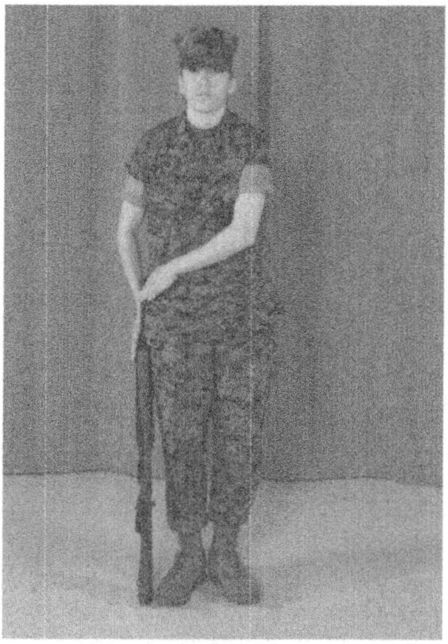

a. Rifle Salute from Order Arms.

b. Rifle Salute from Trail Arms.

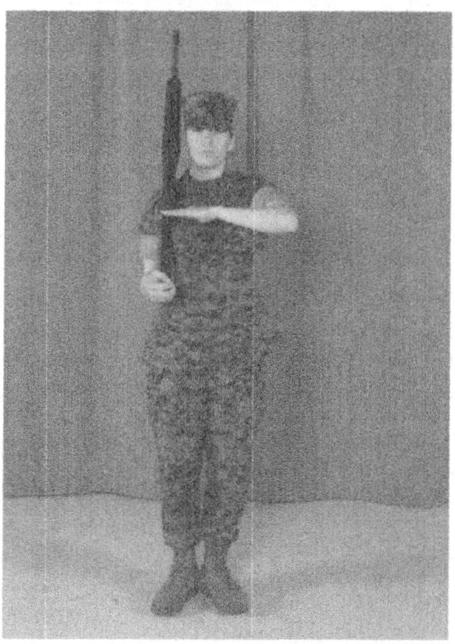

c. Rifle Salute from Right Shoulder Arms.

d. Rifle Salute from Left Shoulder Arms.

Figure 3-17.--Rifle Salutes.

3113. INSPECTION ARMS (WITH AND WITHOUT MAGAZINE)

1. The purpose of this movement is to inspect the rifle when in formation. Order arms to inspection arms is a seven-count movement and is executed when halted at order arms. The command is "**Inspection, ARMS.**"

a. The first two counts are the same as the two-count movement from order arms to port arms. (See figures 3-18b and c.)

b. On the third count, release the grasp of the left hand from the handguard and re-grasp the pistol grip. The fingers are joined and grasping the pistol grip. At the same time, lace the thumb of the left hand over the lower portion of the bolt catch. (See figure 3-18d.)

c. On the fourth count, release the grasp of the right hand from the small of the stock and unlock the charging handle with the thumb and forefinger. The remaining three fingers are joined and placed on line with the forefinger just behind the charging handle. Sharply pull the charging handle to the rearmost position. At the same time, apply pressure to the bolt catch with the thumb of the left hand and lock the bolt to the rear. (See figure 3-18e.)

d. On the fifth count, push the charging handle forward until it is locked in its foremost position and grasp the small of the stock with the right hand. (See figure 3-18f.)

e. On the sixth count, elevate the rifle up and to the left rotating the rifle counterclockwise so that the chamber is visible. The ejection port will be at eye level. The right forearm will touch the stock and the weapon will be at approximately a 45-degree angle. At the same time, turn the head to the left and inspect the chamber to see that it is clean. (See figure 3-18g.)

f. On the seventh count, and after finding the chamber clear or clearing it, lower the rifle rotating it clockwise while turning the hand back to the front. Without loss of motion, release the grasp of the left hand from the pistol grip and re-grasp the handguard resuming the position of port arms. (See figure 3-18h.)

2. To execute inspection arms with the magazine in the weapon, remove the magazine with the left hand and place it between the clothing and the belt (left front) prior to count three.

3. Inspection arms may also be executed from right or left shoulder arms. For the first two counts the rifle is moved from the right (left) shoulder to port arms. The third through seventh counts then executed as described in paragraph 3113.1b through 3113.1f.

a. Start. Count One. Count Two.

b. Count Three. c. Count Four. d. Count Five.

Figure 3-18.--Inspection Arms from Order Arms.

e. Count Six.

f. Count Seven.

Figure 3-18.--Inspection Arms from Order Arms--Continued.

3114. PORT ARMS FROM INSPECTION ARMS (WITH AND WITHOUT MAGAZINE)

1. The purpose of this movement is to terminate inspection arms (rifle bolt locked to the rear). This is a one-count movement, however there are four steps to be accomplished following the preparatory command. It is executed when halted at inspection arms. The command is **"Port, ARMS."** It is the only command that may be given from inspection arms.

a. On the preparatory command of **"Port,"** release the grasp of the left hand from the handguard and re-grasp the weapon with the thumb and fingers, forming a "U" at the magazine well and trigger guard. Press the bolt catch and allow the bolt to go forward. With the fingertips, push upward and close the dust cover. Slide the left hand down and grasp the pistol grip. The fingers are joined and grasping the pistol grip. At the same time, place the thumb on the trigger so the tip of the thumb is outboard.

b. On the command of execution **"ARMS,"** pull the trigger with the thumb of the left hand. Then release the grasp of the left hand from the pistol grip and re-grasp the lower portion of the handguard resuming the position of port arms.

2. If inspection arms is executed with the magazine, it is returned to the weapon immediately after pulling the trigger and before resuming port arms.

3115. FIX AND UNFIX BAYONETS

1. The purpose of these movements is to attach and remove the bayonet while in formation. They are not precision movements, therefore, there are no counts. They are executed when halted at order arms.

2. To attach the bayonet to the rifle the command is "**Fix, BAYONETS.**"

a. On the command of execution, move the muzzle of the rifle to the left front and grasp the barrel with the left hand while keeping the pistol grip and magazine well pointed forward. With the right hand, unsnap the securing strap and withdraw the bayonet. Turn the point skyward and attach the bayonet to the weapon. While engaging the bayonet stud, with the base of the bayonet, grasp the handle and apply downward pressure until a click is heard. Apply limited upward pressure to ensure that the bayonet is seated securely.

b. Snap the scabbard, securing strap, and then resume the position of order arms.

3. To remove the bayonet from the rifle the command is "**Unfix, BAYONETS.**"

a. On the command of execution, move the muzzle to the left and grasp it with the left hand while keeping the pistol grip and magazine well pointed forward. With the right hand unsnap the securing strap. Release the bayonet from the bayonet stud and remove the bayonet from the muzzle. Keeping your eyes on the bayonet point, return it to the scabbard and insert it with the ring facing to the front.

b. Snap the scabbard securing strap and then resume order arms.

3116. AUTHORIZED MANUAL OF ARMS WITH THE M16 RIFLE

1. All of the following movements may be executed while halted. Those marked by an asterisk (*) may be executed while halted or marching forward at quick time.

2. From Order Arms to

a. Port Arms

b. Right Shoulder Arms

c. Left Shoulder Arms

d. Present Arms

e. Inspection Arms

f. Trail Arms

g. Rifle Salute

h. Rest Positions

i. Fix Bayonets

j. Sling Arms

3. <u>From Port Arms to</u>

   a. Right Shoulder Arms *

   b. Left Shoulder Arms *

   c. Order Arms

   d. Present Arms

4. <u>From Right Shoulder Arms to</u>

   a. Left Shoulder Arms *

   b. Port Arms *

   c. Order Arms

   d. Present Arms

   e. Inspection Arms

5. <u>From Left Shoulder Arms to</u>

   a. Right Shoulder Arms *

   b. Port Arms *

   c. Order Arms

   d. Present Arms

   e. Inspection Arms

6. <u>From Present Arms to</u>

   a. Order Arms

   b. Port, Left or Right Shoulder Arms (When a member of a color guard, or when posted as a sentry)

7. From Inspection Arms to Port Arms only.

CHAPTER 3

MANUAL OF ARMS WITH THE M16 RIFLE

SECTION 2: MANUAL OF ARMS WITH A LOOSENED SLING

3200. GENERAL

1. This Section contains the procedures for executing manual of arms movements with the sling loosened so that the rifle may be carried slung from the shoulder. This is known as the sling arms position.

2. All individual or unit drill movements may be executed while at sling arms, except stack arms and unfix and fix bayonets.

3. Remain at sling arms during all rest movements.

4. Stack arms and fix and unfix bayonets (slings loose) are executed from unsling arms (order arms) only.

5. When in formation at sling arms, execute the hand salute on the command **"Present, ARMS."**

3201. SLING ARMS

1. From Order Arms

a. The purpose of this movement is to change from a parade sling to a loosened sling and to then sling the rifle onto the right shoulder. This is not a precision movement; therefore, there are no counts. From the order arms position (with a parade sling) the command for sling arms is **"Sling, ARMS."**

b. On the command of execution, slide the right hand up and grasp the barrel near the flash suppresser. Without loss of motion, raise the rifle vertically to a diagonal position where the butt is in front of the right hip with the muzzle pointing up and the pistol grip to the left. At the same time, grasp the rifle at the handguard just above the slipring with the left hand. The sling is included in the grasp. The fingers are joined. Place the butt on the right hip. Release the grasp of the right hand and with the left hand move the rifle so that it will rest on the inside of the right elbow and cradle it there. (See figure 3-19.) The muzzle points slightly to the right. Release the grasp of the left hand from the handguard and with both hands loosen the sling. After the sling has been loosened, grasp the sling with your left hand and sling the rifle on the right shoulder in the most direct manner. Re-grasp the sling with the right hand. With the exception of the right arm, return to the position of attention. The palm of the right hand is toward the sling. The fingers are joined. The fingers and thumb are wrapped around the sling with the knuckles forward. The wrist and forearm are straight and parallel to the deck. The elbow is holding the rifle in a vertical position and against the body. (See figure 3-20.)

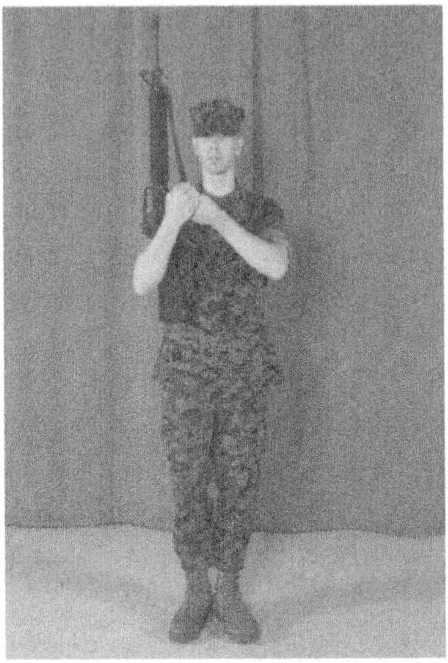

Figure 3-19.--Cradle Rifle Inside Right Elbow.

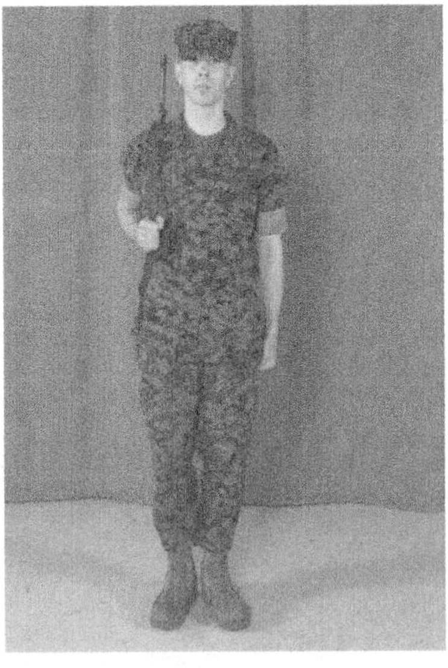

Figure 3-20.--Position of Sling Arms.

2. From Unsling Arms

a. The purpose of this movement is to sling the rifle on the right shoulder when the sling has already been loosened. This is not a precision movement; therefore, there are no counts. From the unsling arms position (order arms with loosened sling), the command for sling arms is "**Sling, ARMS**."

b. On the command of execution, slide the right hand up and grasp the barrel near the flash suppresser. The fingers are joined and wrapped around the barrel with the thumb wrapped around the inboard portion. Without loss of motion, raise the rifle and grasp the sling with the left hand near the upper sling swivel. Release the grasp of the right hand and, with the left hand, sling the rifle over the right shoulder in the most convenient manner. Re-grasp the sling with the right hand. With the exception of the right arm, return to the position of attention. The palm of the right hand is toward the sling. The fingers are joined. The fingers and thumb of the right hand are wrapped around the sling with the knuckles forward. The wrist and forearm are straight and parallel to the deck. The elbow is holding the rifle in a vertical position and against the body.

3202. UNSLING ARMS

1. The purpose of this movement is to take the slung rifle off of the shoulder and move it to the order arms position. This is not a precision movement; therefore, there are no counts. It is executed when halted at sling arms. The command is "**Unsling, ARMS**."

2. On the command of execution, grasp the sling with the left hand in front of the armpit and unsling the rifle from the right shoulder in the most convenient manner. Grasp the rifle at the junction of the barrel and the front sight assembly. Release the grasp of the left hand from the sling and, with the right hand, carry the weapon to the right side until the butt is 3 inches from the deck. The barrel is in a vertical position. At the same time, guide the weapon with the left hand until the right thumb is on the trouser seam. The fingers of the left hand are extended and joined and touching the rifle, near the flash suppresser. The palm of the left hand is toward the rear. The left wrist and forearm are straight and the left elbow is in against the body. Quietly lower the rifle to the deck with the right hand and at the same time return the left hand to the left side at the position of attention.

3203. ADJUST SLINGS

1. The purpose of this movement is to change a loosened sling to parade sling. This is not a precision movement; therefore, there are no counts. The command is "**Adjust, SLINGS**."

2. From unsling arms (order arms); on the command of execution, the rifle is brought to a cradle position inside the right elbow as in the movement from order arms to sling arms. While in this position, the sling is tightened to parade sling. The rifle is then returned to order arms.

3. From sling arms, on the command of execution, grasp the sling with the left hand in front of the armpit and unsling the rifle from the right shoulder in the most convenient manner. Then place the butt on the right hip and cradle the rifle inside the right elbow. Tighten the sling to the parade sling position and automatically assume the position of order arms.

3204. SALUTING AT SLING ARMS

1. The purpose of this movement is to render a salute while at sling arms. It is a two-count movement and is executed when halted at sling arms. The command is "**Present, ARMS.**"

2. On the command of execution and for the count of one, reach across the body with the left hand and grasp the sling just above the right hand. On the second count, release the right hand and execute the hand salute. (See figure 3-21.)

3. To resume order arms, the command is "**Order, ARMS.**" On the command of execution lower the right hand smartly to the right side and re-grasp the sling at the original position. After grasping the sling with the right hand, release the sling with the left hand and return it smartly to the position of attention.

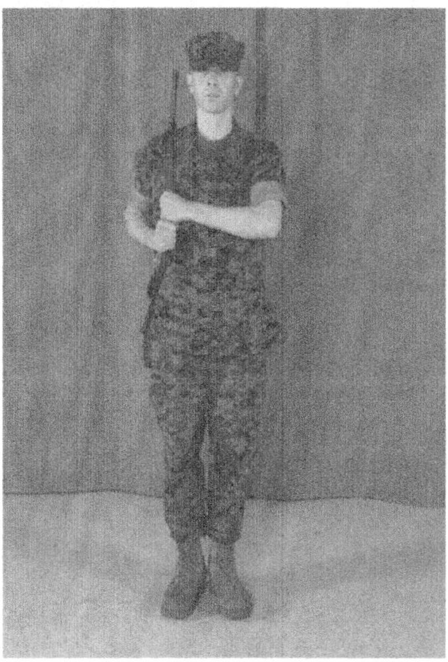

a. Count One.

b. Count Two.

Figure 3-21.--Salute at Sling Arms.

3205. PORT ARMS FROM SLING ARMS

1. The purpose of this movement is take the rifle off of the shoulder and move it to the position of port arms. This is not a precision movement; therefore there are no counts. It is executed when halted at the position of sling arms. The command is "**Port, ARMS**."

2. On the command of execution, reach across the body with the left hand and grasp the sling at the shoulder. Lift the weapon (by the sling), swing it to the front of the body, and grasp the small of the stock with the right hand. Release the sling and re-grasp the weapon just forward of the slipring with the left hand. Keep the elbows into the side with the right forearm horizontal. (See figure 3-22.)

a. Start.

b. Grasping the Sling.

Figure 3-22.--Port Arms from Sling Arms.

c. Weapon off Shoulder.

d. Port Arms.

Figure 3-22.--Port Arms from Sling Arms--Continued.

3. To resume sling arms, the command is "**Sling, ARMS**." On the command of execution, grasp the sling near the upper-sling swivel with the left hand. Release the right hand and swing the weapon back onto the shoulder by inserting the right arm through the sling, immediately resuming the position of sling arms.

3206. INSPECTION ARMS FROM SLING ARMS

1. The purpose of this movement is to inspect the rifle when at sling arms. This is not a precision movement; therefore there are no counts. It is executed when halted at sling arms. The command is "**Inspection, ARMS**."

2. On the command of execution, move the left arm across the body and grasp the rifle by the pistol grip and place the thumb on the lower part of the bolt catch. (See figure 3-23b.) Release the grasp on the sling and, with the left hand and arm supporting the weapon, bring the rifle to a position diagonally across the body, placing the right hand at the small of the stock. (See figure 3-23c.) Unlock the charging handle with the thumb and pull it to the rear. (See figure 3-23d.) Push the charging handle until it is locked in its foremost position and grasp the small of the stock with the right hand. Elevate the rifle up and to the left, at the same time rotating it 90 degrees so that the handgrip is pointing away from the body. At this time, visually inspect the chamber to see that it is clear. (See figure 3-23e.) Return to port arms. (See figure 3-23f.)

a. Start.

b. To the Pistol Grip.

c. Diagonal Across Body.

d. Charging Handle.

e. Inspecting Chamber.

f. Port Arms.

Figure 3-23.--Inspection Arms from Sling Arms.

CHAPTER 3

MANUAL OF ARMS WITH THE M16 RIFLE

SECTION 3: MANUAL OF ARMS MOVEMENTS WITH THE M203

3300. GENERAL

1. This Section contains the procedures for executing manual of arms movements with the M203 Grenade Launcher.

2. When it is necessary to conduct a drill or ceremony involving troops armed with the M203, they will carry the weapon at sling arms. (See figure 3-24.)

3. The M203 will always be rigged with a loosened sling.

4. Troops armed with the M203 will fall in at sling arms and execute all individual drill movements from that position. The only manual of arms movements they will execute are present arms (hand salute) and inspection arms.

5. If stack arms is to be given, troops armed with the M203 will be positioned in ranks so that their weapons are treated as extras on the stacks.

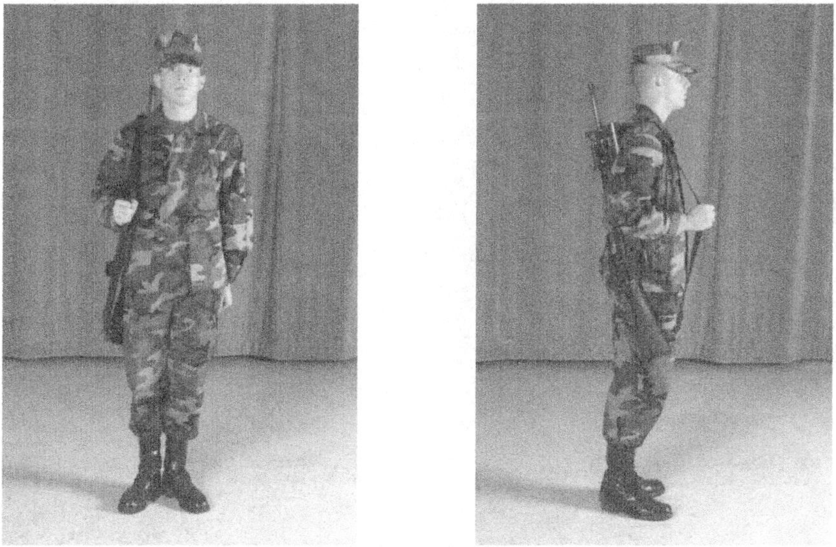

a. Front. b. Right Side.

Figure 3-24.--Sling Arms with the M203.

3301. PRESENT ARMS WITH THE M203

1. When armed with the M203, present arms will be executed in the same manner as the salute while at sling arms with the M16. (See figure 3-21.)

2. Sling arms will be resumed when order arms is given.

3302. INSPECTION ARMS WITH THE M203

1. The purpose of this movement is to inspect the M203 while in formation. This is not a precision movement; therefore, there are no counts. It is executed when halted at sling arms. The command is "**Inspection, ARMS.**"

a. On the command of execution, the initial movements are the same as inspection arms from sling arms with the M16. (See figure 3-23.)

b. When at the inspection arms position for the M16, continue with the M203 by pressing the barrel latch and sliding the barrel up to the barrel stop. Elevate the M203 again, turning the head and eyes, visually inspect the chamber of the barrel. Return to a modified port arms position with the left hand holding the barrel at its full forward position. (See figure 3-25.)

a. Grasp Pistol Grip. b. Grasp Small of Stock. c. Pull Charging Handle. d. Re-grasp Stock. e. Inspect Chamber.

Figure 3-25.--Inspection Arms with the M203.

f. Grasp Barrel Release. g. Open 203 Barrel. h. Left Hand to Port. i. Inspecting M203 Barrel. j. Final Position.

Figure 3-25.--Inspection Arms with the M203--Continued.

2. The command to return to the sling arms position from inspection arms is "**Port, ARMS**."

a. On the preparatory command "**Port,**" slide the barrel down to its closed and latched position. Then continue to close the bolt and dust cover and slide the left hand down to grasp the pistol grip and place the thumb on the rifle trigger as with the M16.

b. On the command of execution "**ARMS**," pull the rifle trigger with the thumb of the left hand and then move it to the trigger of the grenade launcher and pull that trigger. Return to sling arms.

MARINE CORPS DRILL AND CEREMONIES MANUAL

PART I: DRILL

CHAPTER 4

MANUAL OF ARMS WITH THE HANDGUN

FIGURE

MARINE CORPS DRILL AND CEREMONIES MANUAL

PART I: DRILL

CHAPTER 4

MANUAL OF ARMS WITH THE HANDGUN

4000. GENERAL

1. The manual of arms with the handgun is executed with the M9 service pistol. When executing the manual the pistol will be in condition 4 except when executing "**Port, ARMS**" from "**Inspection, ARMS.**" Figure 4-1 shows the nomenclature of the M9 pistol.

2. When in ranks and armed with the pistol, all movements with the exception of inspection arms are executed as if unarmed. The manual of arms for the pistol will be executed with the lanyard attached.

3. The manual of arms for the pistol is not executed in cadence. It is a simple, quick, and safe method of handling the weapon.

4. The manual of arms with the pistol may be executed with the weapon holstered on either the right or left side.

5. When in formation, remain at attention during all rifle manual movements except those listed below.

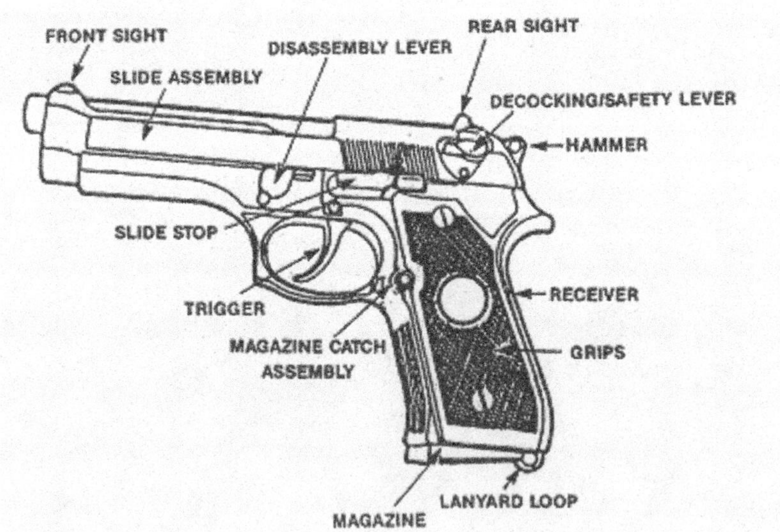

Figure 4-1.--M9 Service Pistol.

4001. PRESENT ARMS WITH PISTOL IN HOLSTER

1. The command is "**Present, ARMS**." It is executed in one count.

2. On the command of execution, "**ARMS**," execute the hand salute.

4002. ORDER ARMS FROM PRESENT ARMS WITH PISTOL IN HOLSTER

1. The command is "**Order, ARMS**." it is executed in one count.

2. On the command of execution, "**ARMS**," smartly return the right hand to its normal position on the right side.

4003. INSPECTION ARMS WITH PISTOL IN HOLSTER

1. The command is "**Inspection, ARMS**." It involves several movements THAT are executed rapidly and smartly without count. It may be executed only when halted at attention with pistol in holster. (See figure 4-2.) Inspection arms is not executed with the pistol as part of the rifle manual except when the unit is formed and dismissed.

NOTE: If holstered on the left side, the opposite hands are used from those described below.

a. Right Handed.

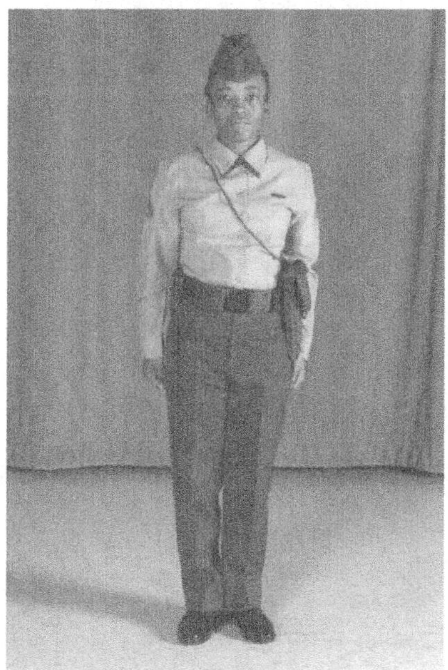

B. Left Handed.

Figure 4-2.--Position of Attention Armed with Pistol.

2. On "**ARMS**," with the right (left) thumb, unfasten the holster flap, grasp the grip and pull the pistol from the holster. Raise the right (left) hand to a position level with and approximately 6 inches in front of the right (left) shoulder. Disconnect the lanyard. The grip should be held between the thumb and last three fingers, forefinger extended and positioned alongside the trigger guard. The muzzle points forward and up at an angle of 30 degrees. (See figure 4-3.)

a. Right Handed.

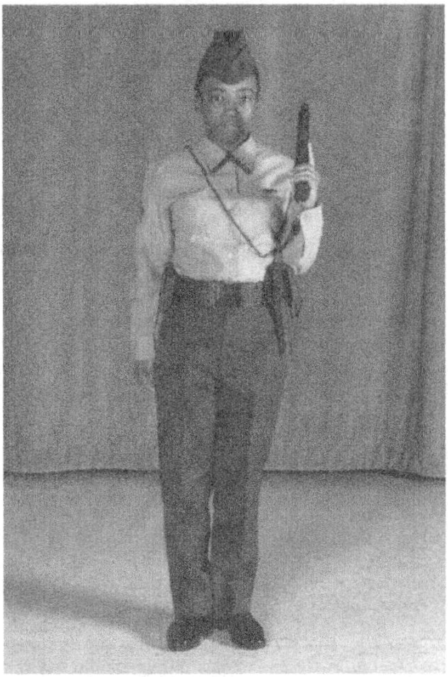

b. Left Handed.

Figure 4-3.--Pistol Removed from the Holster to its Vertical Position.

3. Magazine Removal

a. Right Handed. Without lowering the muzzle or the right hand, turn the pistol handle to the left, look at the pistol, press the magazine catch with the right thumb and remove the magazine with the left hand. (See figure 4-4.) Turn the handle back to the right so that the bottom of the magazine well is to the front. Place the magazine at the left front hip between the pistol belt and outer garment with the magazine follower facing down.

b. Left Handed. Without lowering the muzzle or the left hand, turn the pistol handle to the right, look at the pistol, press the magazine catch with the left forefinger and remove the magazine with the right hand. (See figure 4-5.) Turn the handle back to the left so that the bottom of the magazine well is to the front. Place the magazine at the right front hip between the pistol belt and outer garment with the magazine follower facing down.

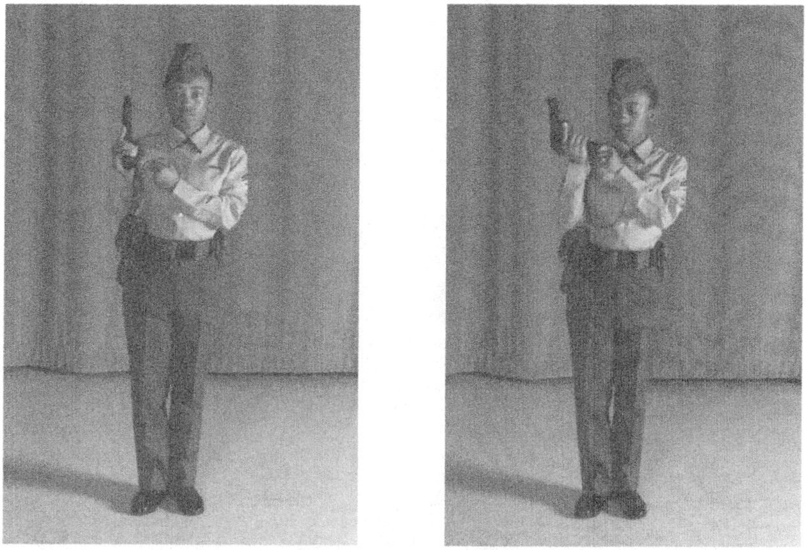

Figure 4-4.--Pistol Rotated while Removing the Magazine (Right Handed).

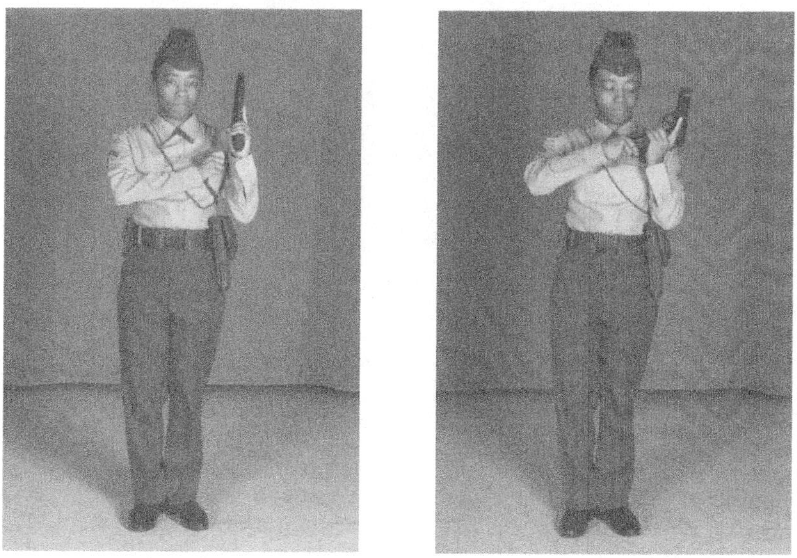

Figure 4-5.--Pistol Rotated while Removing the Magazine (Left Handed).

4. Engaging the Slide Catch/Inspecting the Chamber

a. Right Handed. Without lowering the muzzle or the right hand, grasp the slide with the thumb and fingers of the left hand, thumb on the left side of the slide and pointing downward. Keep the left forearm parallel with the deck. Pull the slide all the way to the rear and engage the slide stop in its notch with the right thumb. (See figure 4-6.) Inspect the chamber to ensure a fully unloaded weapon. Smartly return the left hand to the left side assuming the position of attention. (See figure 4-8.)

Figure 4-6.--Slide to Rear (Right Handed).

b. Left Handed. Without lowering the muzzle or the left hand, grasp the slide with the thumb and fingers of the right hand, thumb on the right side of the slide and pointing downward. Keep the right forearm parallel with the deck. Pull the slide all the way to the rear and engage the slide stop in its notch with the right finger. (See figure 4-7.) Inspect the chamber to ensure a fully unloaded weapon. Smartly return the right hand to the right side assuming the position of attention. (See figure 4-9.)

Figure 4-7.--Slide to Rear (Left Handed).

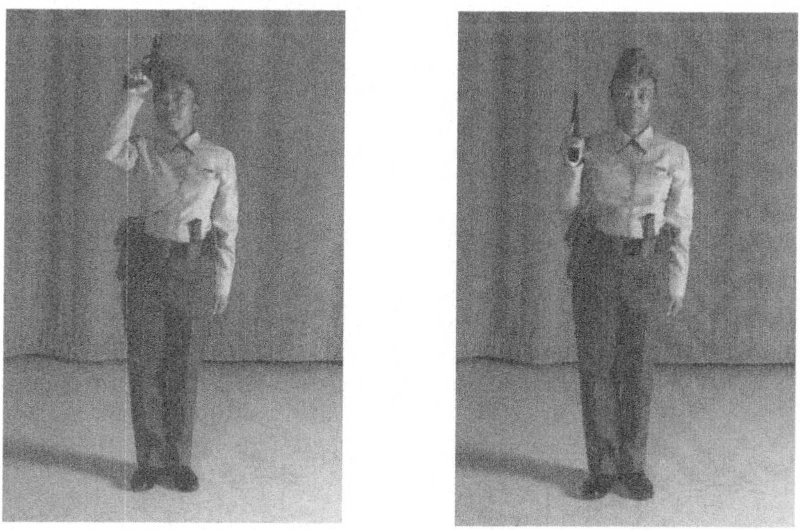

Figure 4-8.--Position of Inspection Arms (Right Handed).

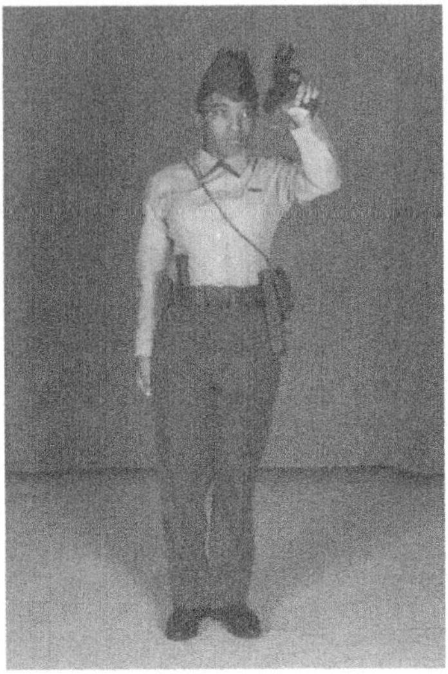

Figure 4-9.--Position of Inspection Arms (Left Handed).

4004. PORT ARMS FROM INSPECTION ARMS (PISTOL RETURNED TO HOLSTER)

1. The command is "**Port, ARMS**" and "**Order, ARMS.**" It is the only command that may be executed from inspection arms. If the pistol is holstered on the left side, the opposite hands are used from those described below.

a. On the command "**Port,**" with the thumb of the right (left) hand release the slide stop. With the muzzle pointing at a 30-degree angle, on the command "**ARMS,**" squeeze the trigger.

NOTE: If the weapon is on safe (condition 3), move the selector switch to fire (condition 1).

b. On the preparatory command "**Order,**" remove the magazine from the pistol belt and insert it back into the magazine well. Re-attach the lanyard to the lanyard ring.

c. On the command "**ARMS,**" return the pistol to the holster and fasten the flap with the right (left) thumb.

d. If the command "**DISMISSED**" is given, pistols will be holstered prior to exiting ranks.

MARINE CORPS DRILL AND CEREMONIES MANUAL

PART I: DRILL

CHAPTER 5

SWORD MANUAL

FIGURE

MARINE CORPS DRILL AND CEREMONIES MANUAL

PART I: DRILL

CHAPTER 5

SWORD MANUAL

5000. GENERAL

1. Authorization for use of the sword.

a. The Marine officers (Mameluke) sword is authorized for use by all Marine commissioned and warrant officers at parades, reviews, and other ceremonies.

b. The Marine noncommissioned officer's sword is authorized for use by all enlisted Marines in the grade of corporal through sergeant major for parades, reviews, and other ceremonies. Marines in the grade of lance corporal through private may be trained in the use of the sword for motivational purposes. However, the use of the noncommissioned officer sword by Marines in the grade of lance corporal and below for parades, reviews or other ceremonies is expressly forbidden.

c. Color guards mounted on horseback may be armed with the noncommissioned officer sword in lieu of rifles or pistols. All members of the color guard need to be a noncommissioned officer and/or staff noncommissioned officer. No other color guards will be armed with swords.

d. Officers, staff noncommissioned officers and noncommissioned officers may be armed with the appropriate sword while filling duty assignments when the commander deems it appropriate to have Marines armed with the sword as a symbol of leadership or authority.

e. A Marine is considered under arms when armed with the sword when the sword is rigged even when not in a duty status. This is because a sword is a ceremonial weapon.

f. The wearing of swords during ceremonies while in the utility uniform is not authorized. Marine Corps uniform regulations prescribe the appropriate uniforms for wear when armed with a sword. However, for instructional purposes, unit leaders may authorize the wear of swords while in the utility uniform during rehearsals and when conducting unit training in sword manual. This provision also applies to sword manual training at all professional military education schools.

2. Drill movements with the sword are made with a fluid, deliberate motion. Trying to "snap" or "whip" the sword will usually result in the execution of the movement seeming to be awkward or ragged. Officers and noncommissioned officers execute the sword manual in the same manner with the following exceptions:

a. The officer's scabbard is rotated to draw or return sword. The Marine officer's scabbard is worn with the convex edge to the rear (see figure 5-1), except when rotated 180-degrees to draw or return sword.

b. The noncommissioned officer's scabbard is attached to a frog that prevents it from rotating. It is worn with the convex edge to the front and remains in that position during draw and return sword. (See figure 5-2.)

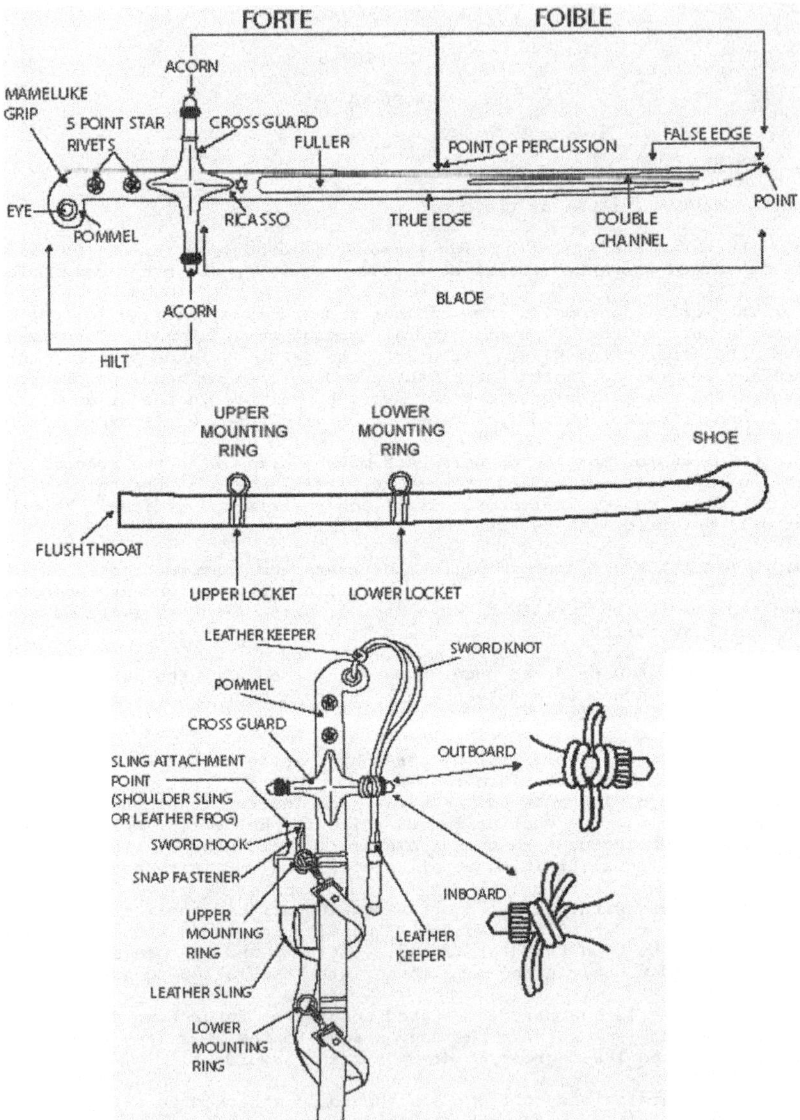

a. Nomenclature.

Figure 5-1.--Marine Officer's Sword.

b. Front.

c. Left Side.

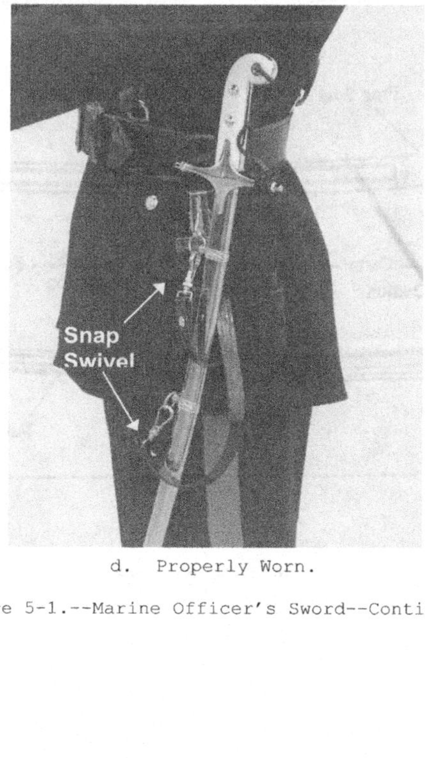

d. Properly Worn.

Figure 5-1.--Marine Officer's Sword--Continued.

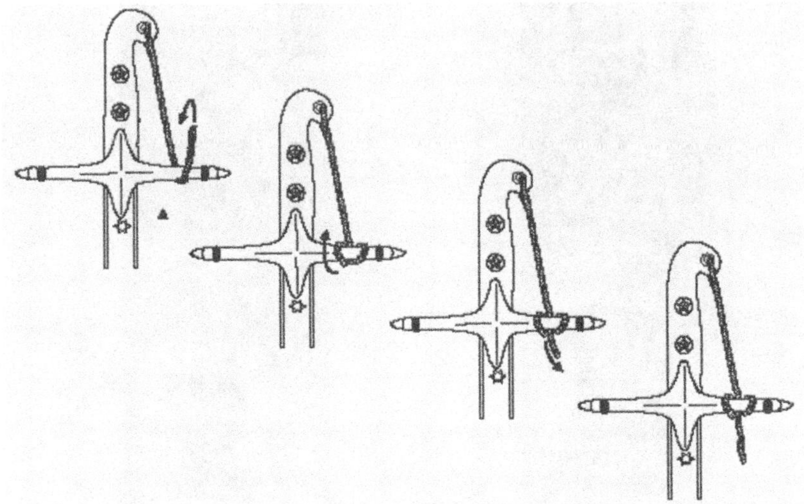

e. Sword Knot.

Figure 5-1.--Marine Officer's Sword--Continued.

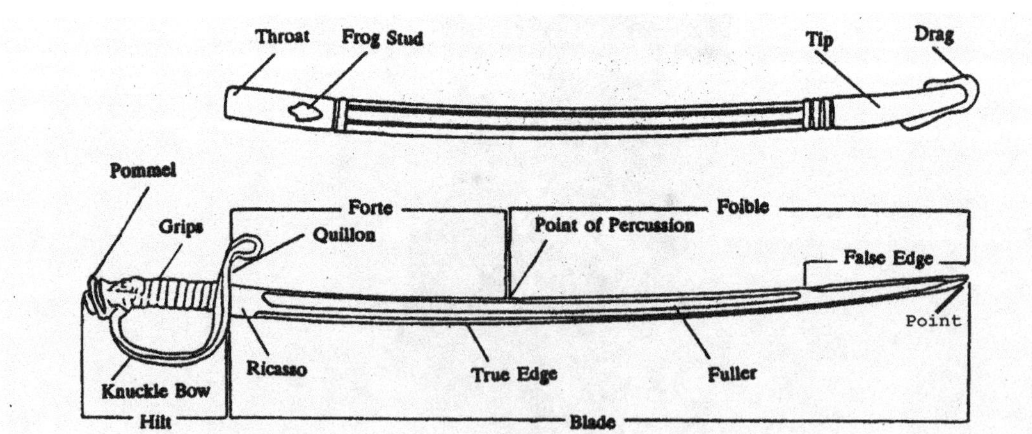

a. Nomenclature.

Figure 5-2.--Noncommissioned Officer's Sword.

b. Front.

c. Left Side.

Figure 5-2.--Noncommissioned Officer's Sword--Continued.

3. Officers and noncommissioned officers draw and return sword, without command, when the commander of their unit does. Unless members of the commander's staff, they execute all other movements of the sword manual on the commander's command to the unit. Staff members execute all other movements on the commander's separate command to his staff. The sword will be drawn with armed troops, except when at ease, rest, route step, at ease march, or when inspecting troops. A unit commander, after being inspected, will return sword prior to accompanying the inspecting party, except during the inspection of an honor guard.

4. Carry Sword is Assumed When

a. Giving commands.

b. Changing position in formation at quick time.

c. Addressing or being addressed by a senior (except when saluting).

d. The preparatory command for any quick time marching movement has been given.

e. Marching at quick time.

f. Company officers and noncommissioned officers go to carry sword when any manual of arms movement has been ordered except parade rest, at ease, rest, present arms, or eyes right (left) (organization staffs excluded).

g. In formation with personnel to your front at normal distance or less (organization staffs excluded), remain at carry sword except during rest or at ease.

h. Commander's staff, to include the staffs of subordinate commanders at regimental/group or larger size units, go to carry sword only when changing position, marching at quick time (e.g., boxing the staff) or on the order of the commander to carry sword prior to marching in review.

5. Present Sword is Assumed When

a. Saluting with the sword.

b. The unit is presented to the colors or any person, or when the "*National Anthem,*" "*To the Colors,*" "*Retreat,*" "*Hail to the Chief,*" memorial "*Taps*" or other musical honors to flag and general officers is played.

c. Executing eyes right (left) while marching past a reviewing officer or stand. If in the interior or rear of a formation, remain at carry sword.

6. Marching with sword at the carry, both arms should swing 6 inches to the front and 3 inches to the rear. Do not hold the scabbard or sling. The exception to this is during eyes right (left) while on the march the right arm is held straight at the side and the sword is kept from moving.

7. Not in formation, keep the sword in its scabbard. Salute by executing the hand salute.

5001. DRAW SWORD

1. The command is "**Draw, SWORD.**"

2. On the Preparatory Command "**Draw**"

a. Officers. Grip the scabbard below the upper brass mounting ring with the left hand. Turn it clockwise 180 degrees. Tilt it forward to form an angle of 45 degrees with the deck. At the same time, reach across the body and grasp the sword grip with the right hand; draw the sword approximately 6 inches from the scabbard until the right wrist and forearm are straight and parallel to the deck. The left hand holds the scabbard against the side. (See figures 5-3a and b.)

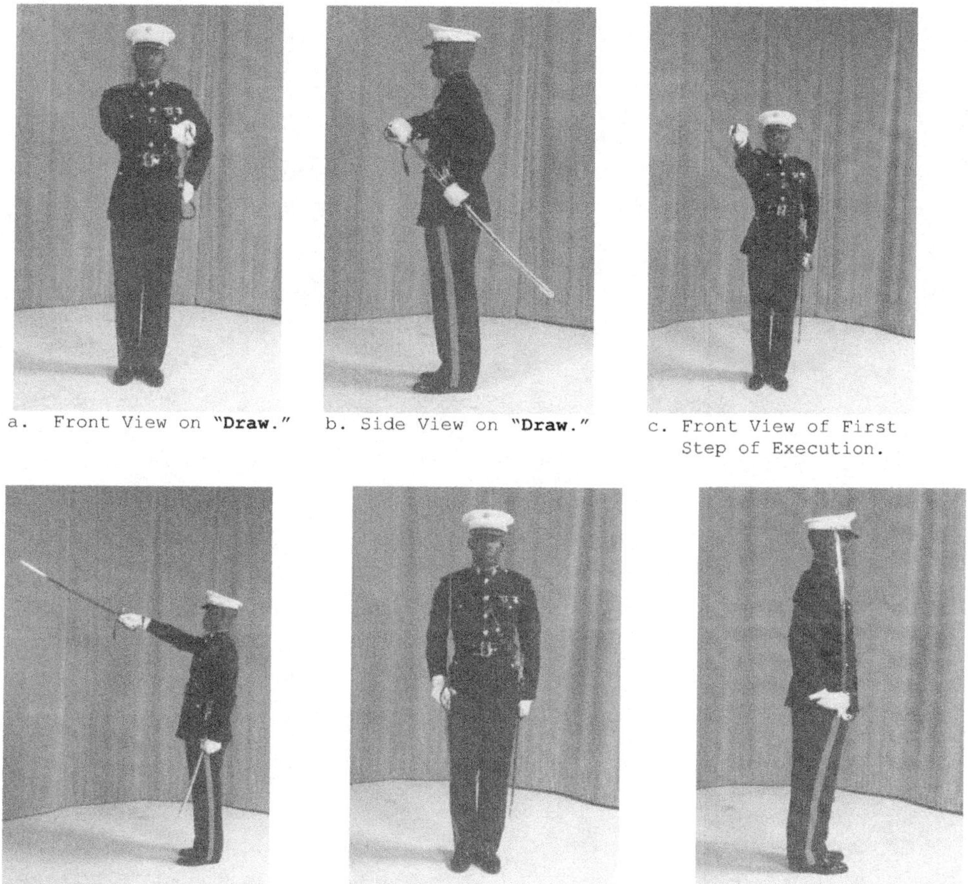

a. Front View on "**Draw.**" b. Side View on "**Draw.**" c. Front View of First Step of Execution.

d. Side View of First Step of Execution. e. Front View of Position of Carry Sword. f. Side View of Carry Sword.

Figure 5-3.--Movements to Draw Sword for Officers.

b. Noncommissioned Officers. Grip the scabbard just below the frog with the left hand. Tilt it forward to form an angle of 45 degrees with the deck. At the same time, reach across the front of the body and grasp the sword grip with the right hand; draw the sword about 10 inches from the scabbard until the right wrist and forearm are straight and parallel to the deck. The left hand holds the scabbard against the side. (See figures 5-4a and b.)

a. Front View on **"Draw."**

b. Side View on **"Draw."**

c. Front View of First Step of Execution.

d. Side View of First Step of Execution.

e. Front View of Position of Carry Sword.

f. Side View of Position of Carry Sword.

Figure 5-4.--Movements to Draw Sword for Noncommissioned Officers.

3. <u>On The Command of Execution **"SWORD"**</u>

a. Draw the sword smartly, raising the right arm to its full extent, directly to the front at an angle of about 45 degrees, the sword in a straight line with the arm, true edge down; drop the left hand to the side. (Officers see figures 5-3c and d; noncommissioned officers see figures 5-4c and d.)

b. Pause for one count.

c. Bring the false edge of the blade against the shoulder seam, blade vertical, back of the grip to the rear, and the arm nearly extended. The right thumb and forefinger embrace the lower part of the grip, with the thumb against the trouser seam, and the remaining fingers joined in a natural curl behind the end of the hilt as if holding a pen or pencil. This is the position of carry sword. (Officers see figures 5-3e and f; noncommissioned officers see figures 5-4e and f.)

5002. PRESENT SWORD FROM CARRY OR ORDER SWORD

1. The command is "**Present, SWORD (ARMS)**." It may be given only when halted at order sword or carry sword. It is executed in two counts.

2. On "**Present,**" raise the right hand to the level of and 6 inches in front of the neck. Keep the thumb on the left side of the grip, wrist slightly bent, and inner forearm against the body. The blade should incline forward at a 30-degree angle from vertical. (Officers see figures 5-5a and b; noncommissioned officers see figures 5-6a and b.)

3. On "**SWORD (ARMS),**" bring the point down smartly, without whipping, to a position 3 inches above the deck and slightly right of the right foot. Straighten the arm so the knuckle bow (noncommissioned officers) or the acorn (officers) is against the trouser seam. The blade is inclined down and to the front with the true edge to the left. The thumb remains on the left side of the grip. (Officers see figures 5-5c and d; noncommissioned officers see figures 5-6c and d.)

4. When at carry or order sword and it becomes necessary to salute without command, execute present sword. After the salute has been returned, go to order sword (then to carry sword if walking).

a. On Command "**Present**" (Front).

b. On Command "**Present**" (Side).

Figure 5-5.--Movements for Present Sword from Carry or Order Sword; Officers.

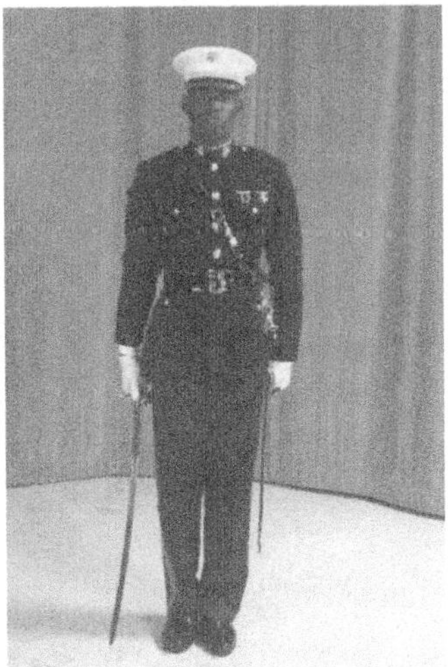

c. Present Sword, Front View.

d. Present Sword, Side View.

Figure 5-5.--Movements for Present Sword from Carry or Order Sword; Officers--Continued.

a. On Command **"Present"** (Front).

b. On Command **"Present"** (Side).

Figure 5-6.--Movements for Present Sword from Carry or Order Sword; Noncommissioned Officers.

c. Present Sword, Front View. d. Present Sword, Side View.

Figure 5-6.--Movements for Present Sword from Carry or Order Sword; Noncommissioned Officers--Continued.

5003. ORDER SWORD FROM PRESENT SWORD

1. The command is "**Order, SWORD (ARMS)**." It is executed in one count.

2. On "**SWORD (ARMS)**," turn the true edge down. In this position, the right arm hangs naturally with the thumb along the trouser seam. The blade slants down to the front with the point 3 inches from the deck. (Officers see figure 5-7, noncommissioned officers see figure 5-8.)

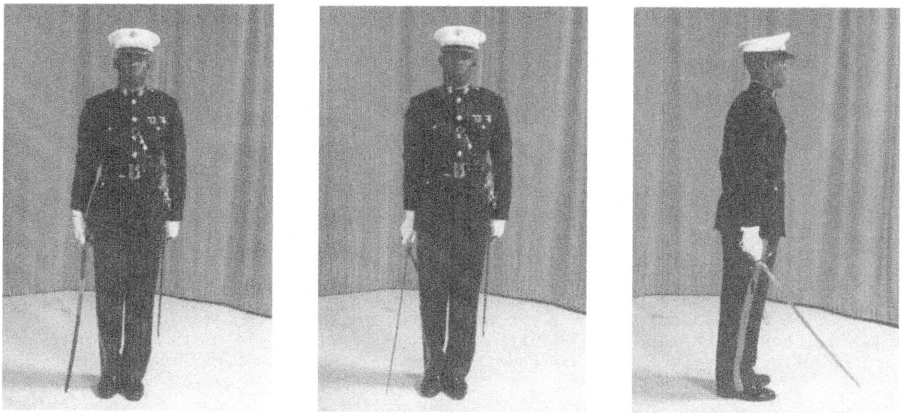

a. Present Sword. b. Order Sword, Front. c. Order Sword, Side.

Figure 5-7.--Order Sword from Present Sword; Officers.

a. Present Sword. b. Order Sword, Front. c. Order Sword, Side.

Figure 5-8.--Order Sword from Present Sword; Noncommissioned Officers.

5004. CARRY SWORD FROM ORDER SWORD OR PRESENT SWORD

1. The command is "**Carry, SWORD**" it may be given only when halted at order sword or present sword. Company officers and noncommissioned officers would also execute carry sword when the unit is given any manual of arms movement (e.g., port, left/right shoulder arms).

2. When at order sword and the command "**Carry, SWORD**" is given. On the command of execution "**SWORD,**" bring the false edge of the blade against the shoulder seam, blade vertical, back of the grip to the rear, and the arm nearly extended. The right thumb and forefinger embrace the lower part of the grip, with the thumb against the trouser seam, and the remaining fingers joined in a natural curl behind the end of the hilt. This is the position of carry sword. (Officers see figures 5-3e and f; noncommissioned officers see figures 5-4e and f.)

3. When at present sword and the command of "**Carry, SWORD**" is given.

a. On the preparatory command of "**Carry,**" turn the true edge down. In this position, the right arm hangs naturally with the thumb along the trouser seam. The blade slants down to the front with the point 3 inches from the deck. (Officers see figures 5-7b and c; noncommissioned officers see figures 5-8b and c.)

b. On the command of execution "**SWORD,**" bring the false edge of the blade against the shoulder seam, blade vertical, back of the grip to the rear, and the arm nearly extended. The right thumb and forefinger embrace the lower part of the grip, with the thumb against the trouser seam, and the remaining fingers joined in a natural curl behind the end of the hilt. This is the position of carry sword. (Officers see figures 5-3e and f; noncommissioned officers see figures 5-4e and f.)

5005. EYES RIGHT (LEFT) FROM CARRY OR ORDER SWORD

1. The command is "**Eyes, RIGHT (LEFT).**" It may be given when halted at order sword or when marching at carry sword. It is executed in two counts.

2. Eyes Right (Left) While Halted

a. On "**Eyes,**" raise the right hand to the level of and 6 inches in front of the neck. Keep the thumb on the left side of the grip, wrist slightly bent, and inner forearm against the body. The blade should incline forward at a 30-degree angle from vertical. This movement is the same as that made on the preparatory command of "**Present.**" (Officers see figure 5-5; noncommissioned officers see figure 5-6.)

b. On "**RIGHT (LEFT),**" bring the point down smartly, without whipping, to a position 3 inches above the deck and slightly right of the right foot. Straighten the arm so the knuckle bow (noncommissioned officers) or the acorn (officers) is against the trouser seam. The blade is slanted down and to the front, with the true edge to the left. The thumb remains on the left side of the grip. At the same time, turn the head and eyes 45 degrees to the right (left). If in extreme right (left) file, continue looking straight ahead. (Officers see figure 5-9; noncommissioned officers see figure 5-10.)

Figure 5-9.--Eyes Right (Left) from Carry or Order Sword; Officers.

Figure 5-10.--Eyes Right (Left) from Carry or Order Sword; Noncommissioned Officers.

3. Order Sword From Eyes Right (Left) (Executed While Halted)

a. The command is "**Ready, FRONT.**" It is executed in one count.

b. On "**FRONT**," turn the true edge of the sword down. At the same time, turn the head and eyes smartly to the front.

4. Eyes Right (Left) While Marching

a. The command is "**Eyes, RIGHT**."

b. The preparatory command "**Eyes**" is given as the right foot strikes the deck.

c. As the left foot strikes the deck raise the right hand to the level of and 6 inches in front of the neck. Keep the thumb on the left side of the grip, wrist slightly bent, and inner forearm against the body. The blade should incline forward at a 30-degree angle from vertical. This movement is the same as that made on the preparatory command of "**Present**." (Officers see figure 5-5a; non-commissioned officers see figure 5-6a.) The left arm continues to swing naturally.

d. The command of execution "**RIGHT**" is given as the right foot strikes the deck.

e. As the left foot strikes the deck bring the point down smartly, without whipping, to a position 3 inches above the deck and slightly right of the right foot. Straighten the arm so the knuckle bow (noncommissioned officers) or the acorn (officers) is against the trouser seam. The blade is slanted down and to the front, with the true edge to the left. The thumb remains on the left side of the grip. At the same time, turn the head and eyes 45 degrees to the right. If in extreme right file, continue looking straight ahead. (Officers see figure 5-9; noncommissioned officers see figure 5-10.) The right arm does not swing. The left arm continues to swing naturally.

f. To execute eyes left while marching, use the above sequence substituting left for right and right for left.

5. Carry Sword from Eyes Right (Left) (Executed When On The March)

a. The command is "**Ready, FRONT**." Execution is begun on the preparatory command.

b. The preparatory command "**Ready**" is given as the left foot strikes the deck.

c. As the right foot strikes the deck turn the true edge of the sword down.

d. The command of execution "**FRONT**" is given as the left foot strikes the deck.

e. As the right foot strikes the deck raise the sword to carry. At the same time, turn the head and eyes to the front.

f. The best way to remember the movements of the sword when executing eyes right and ready front while marching is:

Foot	RIGHT	LEFT	RIGHT	LEFT	RIGHT	LEFT	RIGHT	LEFT
Command	"**Eyes**,	(UP)	**RIGHT**"	(DOWN)	"**Ready**,	(DOWN)	**FRONT**"	(UP)

5006. PARADE REST FROM ORDER SWORD

1. The command is "**Parade, REST.**" This command is normally given from order sword, in which case it is executed in one count.

2. If given when at carry sword, go to order sword on the command of "**Parade.**"

3. On "**REST,**" move the left foot smartly 12 inches to the left. At the same time, lower the point of the sword to the deck, place the left hand behind you, just below the belt. (Officers see figure 5-11, noncommissioned officers see figure 5-12.)

a. Front View. b. Side View.

Figure 5-11.--Parade Rest from Order Sword; Officers.

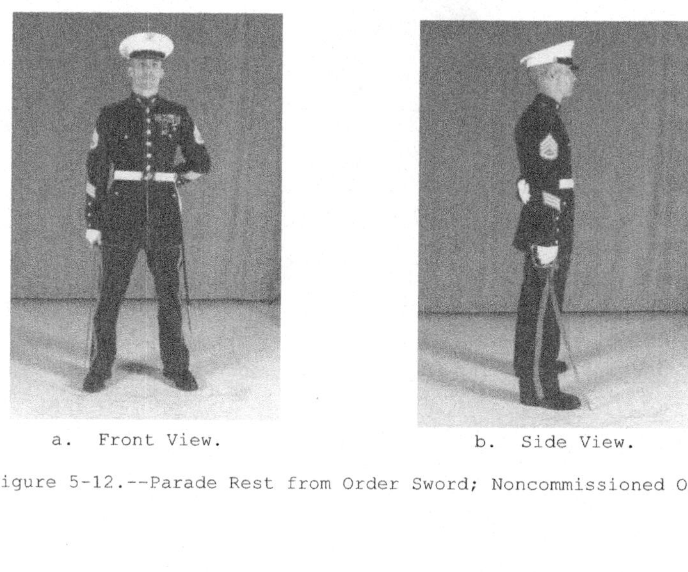

a. Front View. b. Side View.

Figure 5-12.--Parade Rest from Order Sword; Noncommissioned Officers.

4. Order Sword From Parade Rest

a. The command is "**Platoon (Company or Detail), ATTENTION.**" It is executed in one count.

b. On "**ATTENTION,**" bring the left heel smartly against the right and the left hand back to the side. Raise the blade so the point is 3 inches from the deck.

5007. AT EASE FROM ANY POSITION OF THE SWORD

1. The command is "**AT EASE**." It may be given from any position of the sword.

2. On "**AT EASE,**" go to order sword if not already at that position. Then move the left foot 12 inches to the left and lower the point of the sword to the deck. Rest the weight equally on both feet with the legs straight. At the same time, place the left hand behind you. Keep the fingers straight and joined, the palm flat and facing rear. You may relax and, except for your right foot and sword point, move about. Do not talk. (See figure 5-13.)

Figure 5-13.--At Ease.

5008. REST FROM ANY POSITION OF THE SWORD. The command is "**REST**." It is executed in the same manner as at ease, the only difference being that you may talk.

5009. TO RETURN TO ATTENTION

1. The command is "**Platoon (Company or Detail), ATTENTION.**" Execution is begun on the preparatory command.

a. On "**Platoon (Company or Detail)**" go to parade rest.

b. On "**ATTENTION,**" bring the left heel smartly against the right and the left hand to the side. Raise the blade so the point is 3 inches from the deck.

5010. RETURN SWORD FROM CARRY OR ORDER SWORD

1. The command is "**Return, SWORD.**" Execution is begun on the preparatory command.

2. On the preparatory command "**Return.**"

a. Officers raise the right hand and sword to a position 6 inches in front of the neck, as in the first count of present sword. At the same time, grasp the scabbard with the left hand just above the upper brass mounting ring. Tilt it forward and turn it clockwise 180 degrees. The scabbard should form a 45-degree angle with the deck. Then lower the sword point to a position just above the opening of the scabbard. Look down at the opening. Guide the point into the opening with the left thumb and forefinger until the right wrist and forearm is parallel to the deck. Then release the scabbard with the left hand and re-grasp it below the upper brass mounting ring. At the same time, raise the head back to attention. The left hand holds the scabbard against the side. (See figure 5-14.)

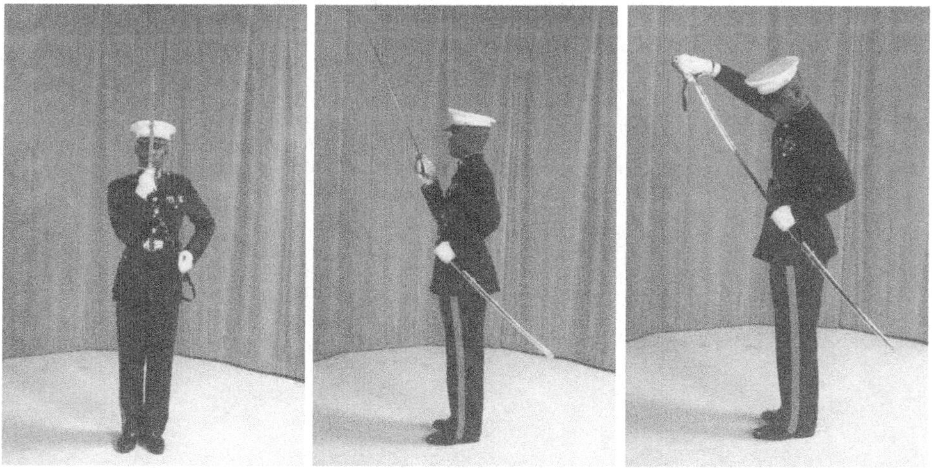

a. Raise Sword and Tilt Scabbard, Front View.

b. Raise Sword and Tilt Scabbard, Side View.

c. Insert Sword into Scabbard, Side View.

Figure 5-14.--Movements for Return Sword from Carry or Order Sword; Officers.

d. Final Position on Preparatory Command "Return," Front View.

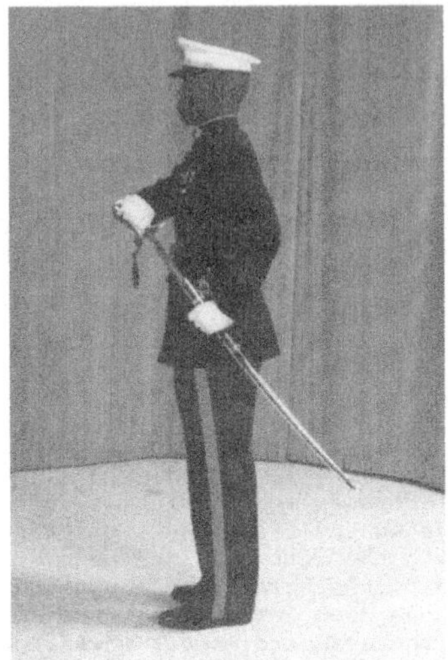

e. Final Position on Preparatory Command "Return," Side View.

Figure 5-14.--Movements for Return Sword from Carry or Order Sword; Officers--Continued.

b. Noncommissioned officers raise the right hand and sword to a position 6 inches in front of the neck as in the first count of present sword. At the same time, position the left thumb and forefinger just below the throat of the scabbard with the remaining fingers lightly grasping the lower front of the frog. Tilt the throat of the scabbard forward with pressure from the left thumb until the scabbard forms a 45-degree angle with the deck. Then lower the sword point to a position just above the opening of the scabbard. Look down at the opening. Guide the point into the opening with the left thumb and forefinger until the right wrist and forearm is parallel to the deck. Then release the scabbard and frog with the left hand and re-grasp the scabbard below the frog. At the same time, raise the head back to attention. The left hand holds the scabbard against the side. (See figure 5-15.)

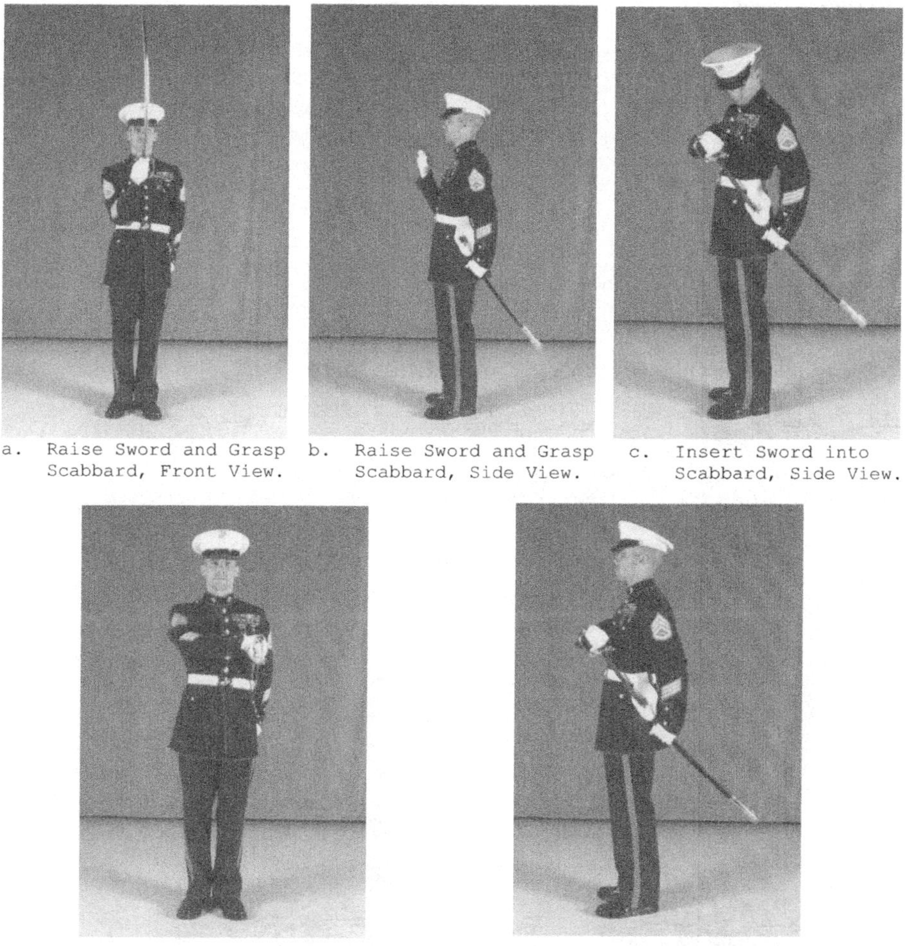

a. Raise Sword and Grasp Scabbard, Front View.

b. Raise Sword and Grasp Scabbard, Side View.

c. Insert Sword into Scabbard, Side View.

d. Final Position on Preparatory Command **"Return,"** Front View.

e. Final Position on Preparatory Command **"Return,"** Side View.

Figure 5-15.--Movements for Return Sword from Carry or Order Sword; Noncommissioned Officers.

c. On the command of execution **"SWORD,"** push smartly down on sword and release the hand grip so it will slide all the way into the scabbard. Then bring the right hand smartly back to the right side. At the same time, release the scabbard with the left hand and assume the position of attention (officers must first turn the scabbard 180 degrees counterclockwise before releasing the scabbard with the left hand).

5011. FUNERAL CARRY

1. This position is assumed when marching for long distances such as when participating in funerals or street parades.

2. The movement to funeral carry may be executed at a pre-arranged point or on command. The command, if given, is "**FUNERAL CARRY**." It is given without rise or inflection of the voice as either foot strikes the deck. This is not a precision movement, therefore, there are no counts. However, there are three steps in executing the movement.

3. At a designated point or on command while at the carry:

a. Reach across the body with the left hand and grasp the sword blade with the thumb and forefinger. The left forearm is parallel to the deck.

b. Change the grasp of the right hand so that the officer's sword grip is grasped between the index and middle fingers, with all fingers curled around the hilt. (See figure 5-16a.) The noncommissioned officer's sword knuckle bow is grasped between the index finger and the thumb; the remaining fingers are joined and curled. (See figure 5-16b.)

c. Release the grasp of the sword blade and return the left hand smartly to the side. Both arms will still swing in the same manner as they did at carry sword.

4. To return to carry sword, the command if given is "**CARRY SWORD**." It will be given without rise or inflection of the voice. At a designated point or on command the above steps will be repeated. However, on step two, the grasp of the sword will change back to the carry sword position.

a. Officer.   b. Noncommissioned Officer.

Figure 5-16.--Funeral Carry.

5012. SWORD MOURNING KNOT

1. The mourning knot will be worn when officers and noncommissioned officers are detailed to military funerals when the sword is prescribed.

2. The mourning knot is a black ribbon of silk or other similar material, 3 inches wide and 27 inches long, each end finished with a small hem. The two flowing ends will be 12 inches long when the band is knotted upon the sword hilt.

3. On the officer's sword, the mourning knot is worn attached to the sword knot. It is formed by doubling the band, passing the free ends around and under the sword knot, immediately below the eye in the pommel, and then passing them back through the bight thus formed. The knot is then drawn taut. (See figure 5-17a.)

4. On the noncommissioned officer's sword, the knot is worn attached to the knuckle bow. It is formed by doubling the band, passing the free ends around the knuckle bow at the pommel, and then back through the bight thus formed and drawn taut. (See figure 5-17b.)

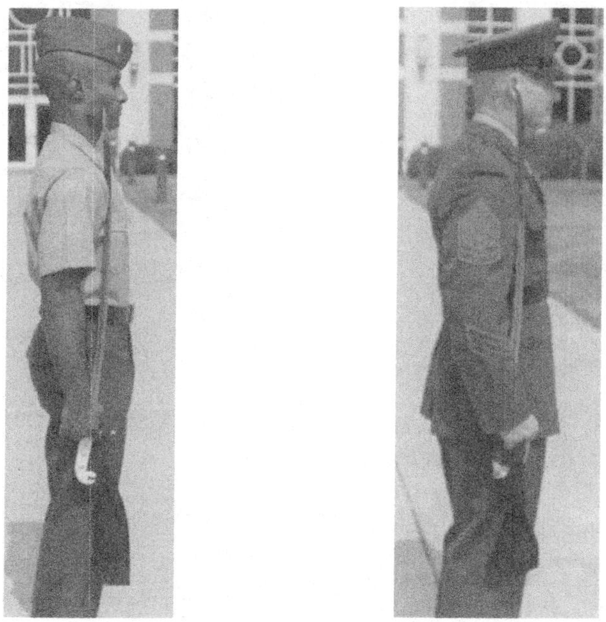

a. Officer. b. Noncommissioned Officer.

Figure 5-17.--The Sword Mourning Knot.

5013. CARRY OF THE SWORD AND SCABBARD UNRIGGED

1. When not in formation and when it is impractical to carry the sword and scabbard rigged at the side (slung or in a frog), the sword sheathed in its scabbard may be carried under the left arm.

2. Place the sword and scabbard under the pit of the left arm with the hilt just to the rear of the left shoulder (knuckle bow up for noncommissioned officers, sword knot up for officers). The drag (noncommissioned officers) or shoe (officers) is to the front with the convex edge of the scabbard up. The left arm is crooked so that the left hand supports the scabbard forward of the shoulder. The left upper arm holds the scabbard against the body. The sword and scabbard slant downward from hilt to drag/shoe. (See figure 5-18.)

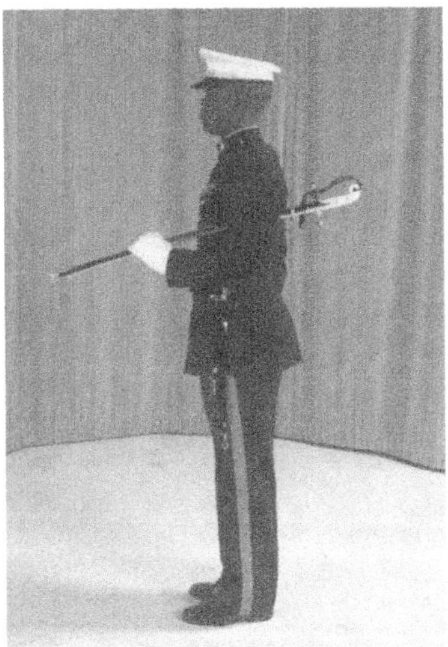

a. Officer.

b. Noncommissioned Officer.

Figure 5-18.--Unrigged Carry.

5014. MARCHING AT DOUBLE TIME WHEN ARMED WITH THE SWORD

1. While marching at double time, hold the sword diagonally across the chest with the true edge to the front. Hold the scabbard with the left hand just below the lower brass mounting ring or the frog. This is not a precision movement but should be executed as follows. (See figure 5-19.)

a. On the preparatory command **"Double Time,"** drop the sword diagonally across the chest with the blade bisecting the head and left shoulder and the true edge to the front. Grasp the scabbard below the lower mounting ring (officer) and below the frog (noncommissioned officer).

b. The movements should be complete prior to the command of execution **"MARCH."**

2. Upon the command **"Quick Time, MARCH"** return to the position of carry sword and release the scabbard when taking the first 30-inch step.

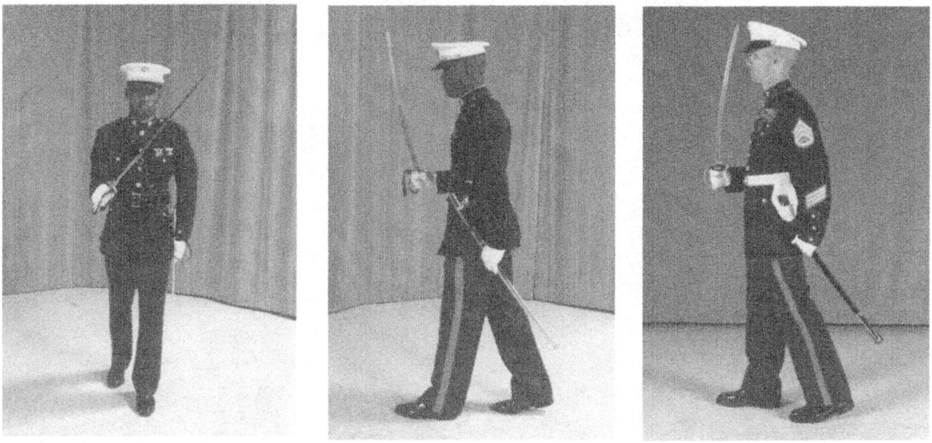

a. Front View, Officer. b. Left Side View, Officer. c. Left Side View, Noncommissioned Officer.

Figure 5-19.--Double Time Armed with a Sword.

5015. PUBLISH THE ORDER

1. When calling roll, reading documents, or publishing orders to a formation, slip the fingers of the left hand between the sword grip and sword knot (officers) or knuckle bow (noncommissioned officers). The sword hangs perpendicular to the deck and is held by the sword knot (officers) or knuckle bow (noncommissioned officers). The true edge is to the rear. Keep the left elbow against the side. Hold the document with both hands. (See figure 5-20.) This is not a precision movement but should be executed as follows.

a. Execute the movement from the position of order sword. Raise and move the sword diagonally across the front of the body until you can grasp the sword knot or knuckle bow with the left hand.

b. Upon completion of the reading reach across with the right hand, re-grasp the sword grip and return the sword to the position of order sword

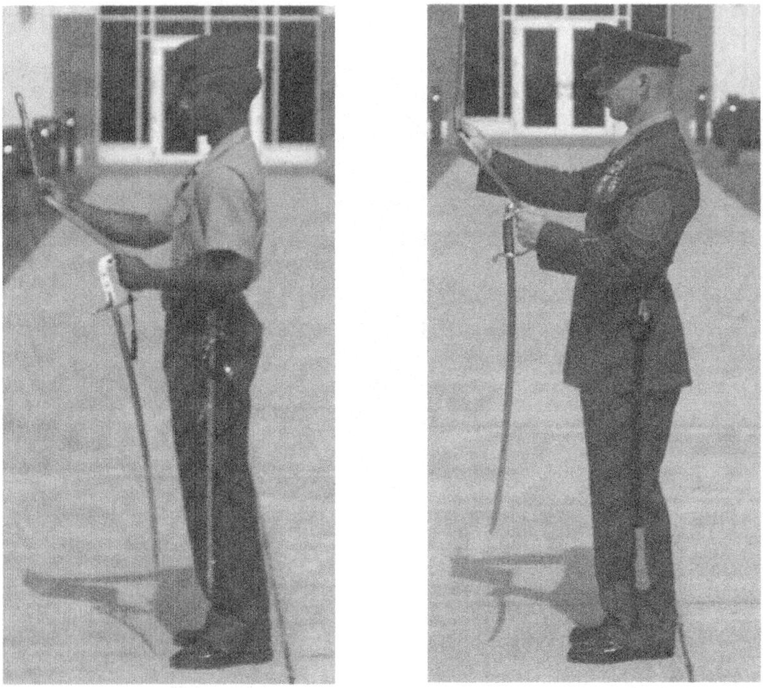

a. Officer.

b. Staff Noncommissioned Officer/ Noncommissioned Officer.

Figure 5-20.--Publish the Order.

5016. FORMING AN ARCH OF SWORDS FOR WEDDINGS

1. The arch of swords for weddings is authorized for commissioned, warrant and staff noncommissioned officers only. The arch of swords ceremony is an old English and American custom, which gives a symbolic pledge of loyalty to the newly married couple from their Marine family. Only the newly married couple is allowed to pass under the arch.

2. The ushers normally form the sword detail, however other officers, warrant or staff noncommissioned officers may be designated as needed. Customarily, six or eight members take part in the ceremony. The usher's form at the bottom of the chapel steps, in two equal ranks, at normal interval, facing each other, with sufficient room between ranks (3 to 4 paces) for the bride and groom to pass. The senior usher is positioned in the left rank furthest from the chapel exit. (See figure 5-21.)

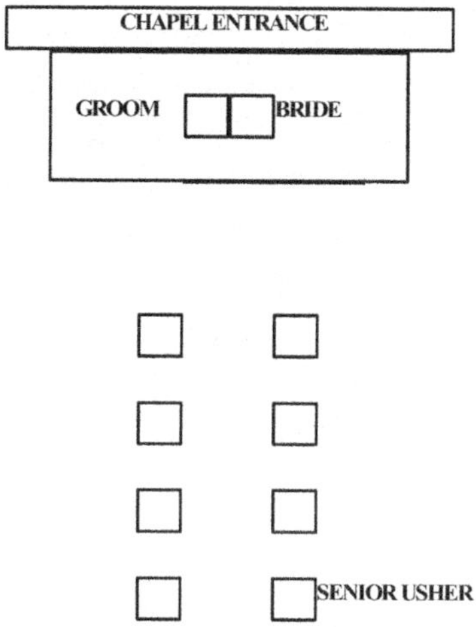

Figure 5-21.--Formation for the Arch of Swords.

3. After the guests have left the chapel to observe the ceremony, the bride and groom exit the chapel and stand just outside the main entrance. The senior usher will then form the arch of swords by giving the command of "**Draw, SWORD**."

a. On the preparatory command of "**Draw**," execute the first count of the movement as described in paragraph 5001.

b. On the command of execution "**SWORD**," the sword is removed from the scabbard in one continuous motion to an angle of 45 degrees, points touching across the aisle formed by the ushers. At this point halt and do not complete the final counts of draw sword.

4. The bride and groom, and only the bride and groom, then pass under the arch. As the newlyweds approach the last two ushers, (e.g., the two farthest from the chapel entrance) they slowly lower their swords to halt the couple. The senior usher then says "Welcome to the Marine Corps Family" or "Welcome aboard Mr./Mrs. __________ (rank of military member can be substituted for Mr./Mrs.), and then announce the new couple to those assembled." The last two ushers then slowly raise their sword back to the arch position and allow the couple to pass.

5. After the newlyweds have cleared the arch the senior usher gives the command of "**Return, SWORD**."

a. On the preparatory command of "**Return**," move the sword to the present position and then to the scabbard as described in paragraph 5010.

b. On the command of execution "**SWORD**," complete the movement of return sword as described in paragraph 5010.

6. The senior usher then dismisses the detail.

7. Though it is traditional that swords are never unsheathed in a chapel, the chaplain may grant permission to form the arch of swords indoors in the event of inclement weather. The ceremony would be performed at the rear of the chapel near or in the vestibule. The ceremony begins after the chaplain has concluded the wedding ceremony with the bride and groom passing under the arch of swords before the guests depart and is performed in the same manner as described above.

5017. PASSING AN UNSHEATHED SWORD

1. At a Marine Corps Birthday cake cutting ceremony or a military wedding reception it is customary to use an officer or noncommissioned officer's sword to cut the birthday or wedding cake.

2. For a Marine Corps cake cutting ceremony the sword is usually placed unsheathed on the cake cart and handed to the commanding general/commanding officer by the senior escort. This is done by laying the sword over the left forearm, cutting edge away from the body, and the hilt towards the commanding general/commanding officer. (See figure 5-23.)

3. At a wedding an officer, warrant or staff noncommissioned officer passes his sword and presents it to his bride, by laying the sword over his left forearm, cutting edge away from the body, hilt towards the bride. (See figure 5-23.) The bride takes the sword and cuts the wedding cake, with the groom's right hand resting over hers on the sword's hilt and with his left arm free to place around his bride. (NOTE: To preclude damaging the sword's blade, ensure it is thoroughly cleaned prior to returning it to the scabbard.)

Figure 5-23.--Passing an Unsheathed Sword.

MARINE CORPS DRILL AND CEREMONIES MANUAL

PART I: DRILL

CHAPTER 6

GUIDON MANUAL

FIGURE

PART I: DRILL

CHAPTER 6

GUIDON MANUAL

6000. GENERAL

1. The guidon is a company or battery identification flag. It is carried on an 8-foot staff at ceremonies and at other times prescribed by the commander. (See figure 6-1.) MCO P10520.3 provides detailed specifications for unit guidons.

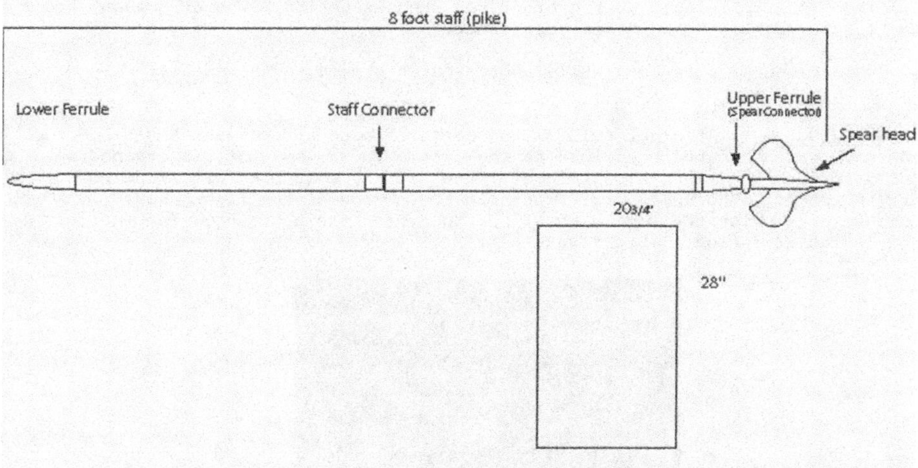

Figure 6-1.--Guidon and Staff.

2. In garrison, the guidon may be displayed at company headquarters between morning and evening colors, except during inclement weather and when carried in formation.

3. In selecting a guidon bearer, the company commander should choose an individual who is outstanding in bearing, appearance, and ability.

4. When acting as guidon bearer, the individual is considered under arms and does not have to carry a weapon. If a weapon is carried, it will be slung or holstered.

5. The guidon is brought to present guidon, parade rest, and order guidon with the company.

6. When at route step or at ease march, the guidon may be carried in either hand. When at order guidon, bring the guidon to carry on the preparatory command for marching in quick time.

7. When at order guidon, the guidon is brought to carry on the preparatory command for any facing or marching movement.

8. The guidon staff should be marked with two thumbtacks. The first tack marks the position of the guidon bearer's right thumb position at order guidon. The second thumbtack is placed 6 inches above the first tack to mark the guidon bearer's right thumb position at carry guidon.

9. The two grasps of the guidon are identified as the "V" grip and the strong grip. The staff is placed in the "V" formed by the thumb and fingers of the right hand. Thumb along the trouser seam with the fingers extended and joined. The strong grip is when the thumb is wrapped around the front of the staff with the fingers wrapped to the rear. (See figure 1-4.)

a. The "V" grip is used at the order and when executing "Present Guidon" and during "Eyes Right/Left" and "Officers Center."

b. The strong grip is used while on the march and facing movements.

6001. ORDER GUIDON. At order guidon, the ferrule rests on the deck touching the outside edge of the right shoe close to the little toe. The staff is grasped in the "V" grip with the right elbow close to the side so that the forearm helps to support the staff. The upper staff rests in the hollow of the right shoulder. The staff is vertical with the flat side of the spearhead facing front. The rest of the body is at attention. (See figure 6-2.)

Figure 6-2.--Order Guidon.

6002. CARRY GUIDON. This is the normal position of the guidon while marching at quick time. It is carried using the strong grip with the ferrule 6 inches above the deck and the flat side of the spearhead facing front. (See figure 6-3.)

Figure 6-3.--Carry Guidon.

6003. DOUBLE TIME WITH THE GUIDON. For marching at double time, on the preparatory command, bring the staff across the body with the spearhead to the left. The right hand grasps the same spot as at carry. The right forearm is level with the deck and the elbow against the body. Grip the staff with the left hand in front of the point where the neck and left shoulder join. The flat side of the spearhead should face front. (See figure 6-4.)

Figure 6-4.--Guidon Carry at the Double Time

6004. CARRY GUIDON FROM ORDER GUIDON. This movement is executed on the preparatory command for any movement that will require the guidon bearer's feet to be moved. Execution is begun on a preparatory command such as "**Forward**," "**About**," etc.

1. On the preparatory command, reach across the body and grasp the staff with the left hand 2 inches above the right.

2. Change the grasp to the strong grip, while raising the staff with the left hand until the ferrule is 6 inches above the deck. Hold the staff vertical with the right hand.

3. Move the left hand smartly back to the side.

6005. ORDER GUIDON FROM CARRY GUIDON. This movement is executed without command at the completion of any movement that required the guidon bearer to come to carry guidon. If there are a series of movements, executed immediately one after the other, order guidon will not be executed until after the final one. Order guidon is executed in one count by allowing the staff to slip through the right hand until the lower ferrule rests on the deck, and change the grasp to the "V" grip.

6006. PRESENT GUIDON FROM ORDER OR CARRY GUIDON. From order or carry guidon, this movement is executed on command. During officers center, the movement is executed after halting, from the carry and without command. Since the right hand is 6 inches lower on the staff at carry than at order, the ferrule will extend farther to the rear when at present guidon from order than it will at present guidon from the carry.

1. From order guidon, the command is **"Present, ARMS."** It is executed in one count.

2. On **"ARMS,"** lower the guidon straight to the front. The right arm is extended until the lower portion of the staff rests in the pit of the right arm and the entire staff is horizontal to the deck. As the staff is lowered rotate the staff to the right so that the sharp edge of the spearhead faces down. Hold the right elbow firmly against the body. (See figure 6-5.)

Figure 6-5.--Present Guidon.

3. From carry guidon the command is **"Eyes, RIGHT (LEFT)."** It is executed in the same manner as from order guidon except that the head and eyes are turned right (left) 45 degrees as the guidon is lowered. (See figure 6-6.)

Figure 6-6.--Eyes Right with the Guidon.

6007. ORDER GUIDON AND CARRY GUIDON FROM PRESENT GUIDON

1. To return to order guidon, the command is "**Order, ARMS.**"

2. On the preparatory command "**Order,**" grasp the staff smartly and audibly with the left hand, palm up, at a point just forward of the right hand. (See figure 6-7.) On "**ARMS,**" with the left hand, pull the guidon staff up and back to the vertical position of carry guidon, pivoting it on the right hand as it is lowered back to the right side. Then let the staff slide straight downward through the right hand until the lower ferrule rests on the deck, the left hand steadies the staff as it slides downward. Cut the left hand smartly back to its normal position at the left side. The flat side of the spearhead is back facing to the front.

Figure 6-7.--Movements to Carry Guidon from Present Guidon.

3. To return to carry guidon the command is "**Ready, FRONT.**"

4. On the preparatory command "**Ready,**" grasp the staff smartly and audibly with the left hand, palm up, at a point just forward of the right hand. On "**FRONT,**" with the left hand, pull the guidon staff up and back to the vertical position of carry guidon, pivoting it on the right hand as it is lowered back to the right side. The flat side of the spearhead facing to the front. Cut the left hand back to the left side and turn the head and eyes smartly to the front.

6008. PRESENT GUIDON FROM CARRY GUIDON AT OFFICERS CENTER. This movement is executed without command, upon halting at officers center during a parade. It is executed in one count as explained in paragraph 6006.2. The movement will be executed on the fourth count after the command "**Officers, HALT.**" For details on executing officers halt refer to paragraphs 17006.9 through 17006.14.

6009. CARRY GUIDON FROM PRESENT GUIDON AT OFFICERS CENTER

1. The command is "**Carry, SWORD.**" "**Ready, TWO**" if the officers are not armed with swords.

2. On the preparatory command **"Carry (Ready),"** grasp the staff with the left hand and on the command of execution **"SWORDS (TWO)"** bring the staff back to the carry guidon position as described in paragraph 6002.

6010. READY GUIDON FROM ORDER OR CARRY GUIDON. This movement is used as a signal for aiding troops in the execution of commands where hearing verbal commands is difficult. When directed by the unit commander, the guidon is brought to ready guidon on specified preparatory commands. Ready guidon may be used as a signal for the execution of any command except present arms and eyes right (left), at which times the guidon must render honors.

1. On the specified preparatory command, bring the left hand across the body to grasp the staff above right hand. Change the grasp of the right hand so that the staff is gripped from the rear. Then raise the right hand to shoulder height, causing the guidon to be raised approximately 3 feet. At the same time, change the grasp of the left hand from above the right hand to just above the lower ferrule. This is the position of ready guidon. (See figure 6-8.)

2. On the command of execution, smartly lower the right hand and staff to the position of carry guidon. At the same time, return the left hand to a position above the right hand and steady the staff as the grasp of the right hand is changed. Return the left hand to the position of attention.

3. Ready guidon may also be executed without command during formation of a unit prior to march on for a review of parade. The guide executes it after the unit has been brought to attention (right shoulder arms if so armed) after the adjutant has commanded, **"Sound, ATTENTION."** Ready guidon is used to signal the adjutant that the unit is ready for his next command.

Figure 6-8.--Ready Guidon.

6011. GUIDON SALUTE FROM CARRY OR ORDER GUIDON

1. The command is "**Guidon (Rifle), SALUTE.**" This movement may be executed on command, when in formation, or individually if the guidon bearer is in route to or from a formation. It is executed in two counts

2. On "**SALUTE,**" move the left hand, palm down, smartly to a position approximately 18 inches above the right hand with the left arm parallel to the deck. Keep the thumb and fingers straight and together. Touch the staff with the first joint of the forefinger. Turn the head and eyes toward the person being saluted. (See figure 6-9.)

Figure 6-9.--Guidon Salute from Carry or Order Guidon.

3. After the salute is returned, or the command "**Order (Ready), ARMS (TWO)**" is given move the left hand smartly back to the left side and look to the front.

6012. RESTS WITH THE GUIDON

1. Parade Rest from Order Guidon

a. The command is "**Parade, REST.**" It is executed in one count. It may be given only when halted at attention.

b. On the command of execution "**REST**," slide the right hand up the staff to shoulder height. The staff remains in place and vertical. Without lost of motion, straighten the right arm so the staff of the guidon tilts forward at 30 degrees. The fingers and thumb of the right hand are wrapped around the staff. The ferrule remains in the same position as at order guidon. The flat side of the spearhead remains facing front. Place the left hand behind the back, just below the belt. Keep the fingers straight and touching. The palm is flat and facing the rear. At the same time, move the left foot smartly 12 inches to the left of the right foot. (See figure 6- 10b.)

a. Front.

b. Side.

Figure 6-10.--Parade Rest with the Guidon.

2. <u>At Ease from Any Position of the Guidon</u>

a. The command is "**<u>AT EASE</u>**." It is executed in two counts. It may be given only when halted at attention.

b. On the command, and for count one, execute parade rest as described in paragraph 6012.1b.

c. On the second count, you may relax and, except for your right foot and ferrule, move about. Do not talk.

3. Rest from Any Position of the Guidon. The command is "**REST**." It is executed in the same manner as at ease the only exceptions being that you may talk and hold the staff in either hand.

4. Fall Out from Order Guidon. The command is "**FALL OUT**." It may be given only from order guidon. Upon the command, leave your position in ranks but remain nearby.

5. Order Guidon from Rest Positions

a. From parade rest, at ease, and rest, the command is "**Platoon (Company or Battalion), ATTENTION.**" On the preparatory command, go to parade rest, if not already in that position. At the command "**ATTENTION,**" go smartly to order guidon.

b. From fall out, the command is "**FALL IN**." Upon the command, go back to your position in formation and stand at attention at order guidon.